METRIC PLANES AND METRIC VECTOR SPACES

METRIC PLANES AND METRIC VECTOR SPACES

ROLF LINGENBERG
University of Karlsruhe

A WILEY-INTERSCIENCE PUBLICATION

JOHN WILEY & SONS, New York · Chichester · Brisbane · Toronto

Copyright © 1979 by John Wiley & Sons, Inc.

All rights reserved. Published simultaneously in Canada.

Reproduction or translation of any part of this work
beyond that permitted by Sections 107 or 108 of the
1976 United States Copyright Act without the permission
of the copyright owner is unlawful. Requests for
permission or further information should be addressed to
the Permissions Department, John Wiley & Sons, Inc.

Library of Congress Cataloging in Publication Data

Lingenberg, Rolf.
  Metric planes and metric vector spaces.

  (Pure and applied mathematics)
  "A Wiley-Interscience publication."
  Bibliography: p.
  Includes index.
  1. Geometry, Non-Euclidean. 2. Geometry, Plane.
  3. Vector spaces. I. Title.

QA685.L72 516'.9 78-21906
ISBN 0-471-04901-8

Printed in the United States of America

10 9 8 7 6 5 4 3 2 1

Math.

Sep.

PREFACE

This book is devoted to a certain domain of plane geometry for which both incidence and metric concepts, such as orthogonality or reflections, are defined.

The theory developed can be regarded as

(a) a purely geometric theory based on the concept of incidence structures with orthogonality or with reflections, mainly as a treatment of Euclidean and non-Euclidean planes and certain subplanes of these planes,

(b) a theory of three-dimensional metric vector spaces with their natural geometric interpretation,

(c) a theory of special types of S-groups and their group planes.

These areas of plane geometry, (a) to (c), are only different representations of the same theory, and one of the most important tasks of the book is to verify these relations.

The first point of view follows the tradition of developing plane metric geometry in an axiomatic treatment. D. Hilbert's foundation of the real Euclidean geometry is the most familiar example of such an axiomatic treatment. Although we start with a very general definition of a metric plane (Chapters 2 and 3), this is done in anticipation of further generalizations, and we are mainly interested in a natural generalization of the concept of classical Euclidean and non-Euclidean planes and in subplanes of these general Euclidean and non-Euclidean planes. In these generalizations the coordinate field of Euclidean and non-Euclidean planes can be the field of the reals but also any other field, even finite fields or fields of characteristic 2. The non-Euclidean planes considered in the book are divided into elliptic, hyperbolic-metric, Minkowskian,* and Strubecker

---

* More precisely, we should say Lorentz-Minkowskian planes, to distinguish the planes defined here from other "Minkowskian" geometry such as the geometry of a convex domain in the real Euclidean plane.

planes.* Adding the Euclidean planes, we obtain five types of complete planes (Chapter 6). In studying subplanes of these planes we restrict ourselves to those cases for which a characterization as a subplane of a complete plane can be given by utilizing certain of their properties without resorting to the embedding into a complete plane (Chapter 7). This part of the theory developed in this book can be regarded as the most general treatment of absolute geometry, whose foundation had been laid in Hilbert's *Grundlagen der Geometrie* and built upon by many other geometers.

The second approach is related to the analytic–algebraic treatment of geometry, based on the method of formulating and solving geometric problems in the language of linear algebra, and thus in the language of vector spaces. Metric geometry then corresponds to the theory of metric vector spaces, that is, pairs (V, Q) consisting of a vector space V and a quadratic form Q (Chapter 1). The case of a metric plane corresponds to the case of a three-dimensional metric vector space, according to the general fact that a projective space of dimension n corresponds to a vector space of dimension $n + 1$. One of the main theorems in this book states that up to isomorphisms all five complete planes mentioned here are all the metric planes over metric vector spaces (V, Q) such that $\dim V^{\perp} \leq 2$ (the case $\dim V^{\perp} = 3$ is not an interesting one).

The last point of view arises from a method used in recent years to describe and study plane metric geometry. This method consists in utilizing the calculus of reflections systematically, and in starting from the fact that lines can be replaced by the reflection in the lines and that geometric relations such as incidence and orthogonality can be expressed by relations in the group generated by the reflections. Thus not only the geometric structures themselves, but also and mainly their groups of automorphisms generated by reflections, come under consideration. The group of motions (i.e., the group of automorphisms) of any Euclidean or non-Euclidean plane over an arbitrary coordinate field (even of characteristic 2) can be regarded as an S-group, and the corresponding S-group plane is isomorphic to the original plane. Here an S-group is a pair (G, S) consisting of a group G and a subset S of G which generates G, containing only involutions and in addition satisfying a certain property denoted Axiom S (Chapter 4). We can associate a group plane with the S-group in the familiar canonical way: lines are the elements of S: three lines a, b, and c are concurrent if the product abc is an involution; and lines a and b are orthogonal if the product ab is an involution. Axiom S may be interpreted geometrically as the three-reflection

* Or "parabolic" planes. Compare references 32 and 47.

theorem. It should be emphasized that E. Sperner was the first to introduce Axiom S in this form.

The systematic use of reflections in the theory of metric planes is due to J. Hjelmslev, and F. Bachmann has brought this method to an outstanding fruition in his book *Aufbau der Geometrie aus dem Spiegelungsbegriff.** Personally I am very much indebted to the work of F. Bachmann.

Because finite geometric structures are of special interest today, Chapter 8 deals with finite S-groups and proves that finite Euclidean, finite hyperbolic-metric, and finite Minkowskian planes can be characterized by very weak assumptions on S-group planes. As an immediate consequence we obtain that finite elliptic planes do not exist.

The key of characterizing all the Euclidean and non-Euclidean planes given in this book is the concept of Δ-connected points. Therefore Chapter 5 scrutinizes consequences of the existence of Δ-connected points.

The Appendix collects all the definitions and results from affine and projective geometry and from group theory which are needed in the text. Moreover, the concept of projective reflection groups is given and briefly discussed.

This book arose from a lecture given at the Conference on Foundations of Geometry, held at the University of Toronto in 1974. I thank Professor Dr. P. Scherk very much for inviting me to that lecture.

I would like to express special thanks to Dr. G. Gunther, who carefully read the manuscript and improved the presentation. Also I am very much indebted to R. K. Becker, Dr. V. Drumm, and Dr. A. Häussler for proofreading and for suggesting many improvements. Moreover, I would like to give thanks to Ms. G. Lösch for typing and preparing the manuscript and for drawing the figures with special care.

R. LINGENBERG

Karlsruhe, Germany
December 1978

---

* For a more detailed report on the historical development, see reference 2.

CONTENTS

METRIC PLANES AND
METRIC VECTOR SPACES

1
METRIC VECTOR SPACES

1. BASIC DEFINITIONS

Let (V, K) be a vector space over the field K. We denote the subspace of V generated by the vectors $A, B, C, \ldots$, by $\langle A, B, C, \ldots \rangle$.

A mapping

$$Q : \begin{cases} V \to K \\ X \mapsto Q(X) \end{cases}$$

is called a *quadratic form* on V if it satisfies the following properties:

(Q1) $Q(aX) = a^2 Q(X)$ *for all* $a \in K$ *and* $X \in V$.

(Q2)   *The mapping*

$$g : \begin{cases} V \times V \to K \\ (X, Y) \mapsto Q(X+Y) - Q(X) - Q(Y) \end{cases}$$

      is a bilinear form.

This bilinear form is called the form belonging to Q, sometimes denoted g_Q. Obviously, from (Q2) we see that g_Q is symmetric. Also from (Q2) we can easily deduce that

$$g(X, X) = 2Q(X) \quad \text{for all} \quad X \in V \tag{1}$$

The mapping that associates 0 to every vector X is a quadratic form Q on V. We write $Q = 0$ and call Q the *zero form*.

Suppose now that char $K \neq 2$. Let f be any symmetric bilinear form, and define

$$Q : \begin{cases} V \to K \\ X \mapsto f(X, X) \end{cases}$$

1

Then Q is a quadratic form and $g := 2f$ is the form belonging to Q. We sometimes write Q_f instead of Q.

We note the following rules:

$$\left.\begin{array}{l} Q(A+B) = Q(A) + Q(B) + g(A, B) \\ Q(A-B) = Q(A) + Q(B) - g(A, B) \end{array}\right\} \text{ for all } \quad A, B \in V \qquad (2)$$

The first equation follows immediately from the definition of a quadratic form. Combining this equation with properties (Q1) and (Q2) immediately yields $Q(A-B) = Q(A+(-1)B) = Q(A) + Q((-1)B) + g(A, (-1)B) = Q(A) + Q(B) - g(A, B)$.

Let (V, K) be a vector space, Q a quadratic form, and g the bilinear form belonging to Q. Then the pair (V, Q) is called a *metric vector space*. [We note here that if char $K \neq 2$ and $Q = Q_f$, then we put $(V, f) := (V, Q_f)$.] A vector X is called *isotropic** if $Q(X) = 0$, and two vectors X, Y are called *orthogonal* if $g(X, Y) = 0$. If X is isotropic, then X is orthogonal to itself by equation 1.

If T is any subspace of V, we define

$$T^{\perp} := \{X \in V \mid g(X, Y) = 0 \quad \text{for all} \quad Y \in T\}$$

Clearly, $T^{\perp}$ is a subspace, called the *subspace orthogonal to T*. The set rad $(V, Q) := \{X \mid X \in V^{\perp}, \text{ and } Q(X) = 0\}$ is also a subspace, called the *radical* of (V, Q).

If char $K \neq 2$, then rad $(V, Q) = V^{\perp}$, since $X \in V^{\perp}$ implies that $g(X, X) = 0$, and thus, by equation 1, $Q(X) = 0$.

We now can prove

LEMMA 1.1.   *Let* (V, Q) *be a three-dimensional metric vector space and* $A \in V$. *Then*

(a)   *If* $A \in V^{\perp}$,   *then* $\langle A \rangle^{\perp} = V$.
(b)   *If* $A \notin V^{\perp}$,   *then*   *dim* $\langle A \rangle^{\perp} = 2$.
(c)   *If* $g_Q(A, A) \neq 0$,   *then* $V = \langle A \rangle \oplus \langle A \rangle^{\perp}$.

PROOF.   Clearly, (a) is true.

   (b): We consider the mapping

$$\alpha : \begin{cases} V \to K \\ X \mapsto g_Q(A, X) \end{cases}$$

---

* Vectors X for which $Q(X) = 0$ are often called *singular*, while the term *isotropic* is reserved for X satisfying $g(X, X) = 0$. If char $K \neq 2$, $Q(X) = 0$ if and only if $g(X, X) = 0$, hence this distinction vanishes. If char $K = 2$, we saw that $g(X, X) = 0$ for all X, and thus all vectors are isotropic in this sense. As only the vectors X satisfying $Q(X) = 0$ are exceptional in orthogonal geometry, we call them isotropic.

Clearly, α is a linear form. Since $A \notin V^{\perp}$, there exists some vector $X \in V$ such that $g_Q(A, X) \neq 0$. Hence Im $\alpha \neq O$,* and thus dim (Im α) = 1. We conclude that dim $\langle A \rangle^{\perp} = $ dim (ker α) = dim $V -$ dim (Im α) = 3 - 1 = 2.

(c): We recall the dimension theorem for subspaces:

$$\dim (T + T') = \dim T + \dim T' - \dim (T \cap T')$$

Since $g_Q(A, A) \neq 0$, we know that $A \notin \langle A \rangle^{\perp}$ and therefore $\langle A \rangle \cap \langle A \rangle^{\perp} = O$. Hence we can apply the dimension theorem to $\langle A \rangle$ and $\langle A \rangle^{\perp}$ to obtain dim $(\langle A \rangle + \langle A \rangle^{\perp}) = 3$. Thus $\langle A \rangle + \langle A \rangle^{\perp} = V$ and, since $\langle A \rangle \cap \langle A \rangle^{\perp} = O$, we even have $\langle A \rangle \oplus \langle A \rangle^{\perp} = V$, as required. ∎

2. ISOTROPIC SUBSPACES AND WITT-INDEX

Let (V, Q) be a metric vector space. A subset M of V is called *isotropic* if any vector in M is isotropic. If M is isotropic, we write $Q(M) = 0$ instead of "$Q(X) = 0$ for all X in M". Especially, if M is a subspace, we call M an *isotropic subspace*.

Obviously, the subspace rad (V, Q) is an isotropic subspace.

LEMMA 1.2. *Let* A, B *be any two isotropic vectors in a metric space* (V, Q) *and* $g := g_Q$. *Then*

$$A + B \text{ is isotropic} \Leftrightarrow g(A, B) = 0 \Leftrightarrow \langle A, B \rangle \text{ is isotropic}$$

PROOF. If $A + B$ is isotropic, then, by definition of the bilinear form g, $g(A, B) = Q(A + B) - Q(A) - Q(B) = 0$. Now let $g(A, B) = 0$. Then for any two elements x, y in the underlying field of V we obtain $Q(xA + yB) = x^2 Q(A) + y^2 Q(B) + xyg(A, B) = 0$, hence $\langle A, B \rangle$ is isotropic. If $\langle A, B \rangle$ is isotropic, then in particular $A + B$ is isotropic. This proves Lemma 1.2. ∎

COROLLARY 1.3. *Let* (V, Q) *be any metric vector space and* T *any subspace of* V. *If* T *is isotropic, then* $T + $ rad (V, Q) *is isotropic.*

PROOF. Let $Z = X + Y$ be any vector in $T + $ rad (V, Q) and $X \in T$ and $Y \in $ rad (V, Q). Then $Q(Z) = Q(X) + Q(Y) + g(X, Y) = 0$. Hence $T + $ rad (V, Q) is isotropic. ∎

LEMMA 1.4. *Any nonisotropic two-dimensional subspace of a metric vector space contains at most two distinct one-dimensional isotropic subspaces.*

---

* For the sake of brevity we identify a set consisting of exactly one element with that element.

PROOF.   We assume that there are three distinct one-dimensional isotropic subspaces in a two-dimensional subspace T. Without loss of generality we may assume $\langle A \rangle$, $\langle B \rangle$, $\langle A + B \rangle$ are the three distinct isotropic subspaces in T. Then $T = \langle A, B \rangle$ and, by Lemma 1.2, T is isotropic. This proves Lemma 1.4.   ■

LEMMA 1.5.   *Let* (V, Q) *be a finite dimensional metric vector space and* T *an isotropic subspace of maximal dimension. Then rad* (V, Q) ⊂ T.

PROOF.   By Corollary 1.3, $T + \mathrm{rad}\,(V, Q)$ is isotropic and dim $T \le$ dim $(T + \mathrm{rad}\,(V, Q)) \le \dim T$, since dim T is the maximal dimension of an isotropic subspace. This yields $T = T + \mathrm{rad}\,(V, Q)$ and therefore $\mathrm{rad}\,(V, Q) \subset T$, as required.   ■

Let (V, Q) be a finite-dimensional metric vector space, and let T be an isotropic subspace of maximal dimension; then the number dim T − dim rad (V, Q) is called the *Witt-index* of the metric vector space (V, Q), and is denoted by Ind (V, Q).

LEMMA 1.6.   *Let* (V, Q) *be a finite-dimensional metric vector space. Then Ind* (V, Q) = 0 *if and only if the following property holds:*

$$X \text{ isotropic} \Rightarrow X \in \mathrm{rad}\,(V, Q) \qquad\qquad (*)$$

PROOF.   (A) Let Ind $(V, Q) = 0$. Then we have for any isotropic subspace T of maximal dimension dim $T = $ dim rad (V, Q). If X is any isotropic vector $\ne O$ [if $X = O$, then $X \in \mathrm{rad}\,(V, Q)$], then by Corollary 1.3 $\langle X \rangle + \mathrm{rad}\,(V, Q)$ is isotropic and dim rad $(V, Q) \le \dim(\langle X \rangle + \mathrm{rad}\,(V, Q)) \le \dim \mathrm{rad}\,(V, Q)$, since dim rad (V, Q) is the maximal dimension of an isotropic subspace. Hence we obtain $\langle X \rangle + \mathrm{rad}\,(V, Q) = \mathrm{rad}\,(V, Q)$ and $X \in \mathrm{rad}\,(V, Q)$, as required. Thus $(*)$ holds

(B) Let $(*)$ be valid. Let T be any isotropic subspace of maximal dimension. By $(*)$ we obtain $T \subset \mathrm{rad}\,(V, Q)$, hence, by Lemma 1.5, $T = \mathrm{rad}\,(V, Q)$ and Ind $(V, Q) = $ dim $T − $ dim rad $(V, Q) = 0$, as required.

Thus Lemma 1.6 is proved.   ■

LEMMA 1.7.   *Let* (V, Q) *be a three-dimensional metric vector space and* Q ≠ 0. *Then dim rad* (V, Q) + *Ind* (V, Q) ≤ 2 *and Ind* (V, Q) ≤ 1.

PROOF.   Let T be any isotropic subspace of maximal dimension. From $Q \ne 0$ we deduce $T \ne V$, hence dim $T \le 2$. Therefore Ind $(V, Q) + $ dim rad $(V, Q) = $ dim $T \le 2$ by definition.

If dim $T \le 1$, then, by definition, Ind $(V, Q) \le 1$. Now let dim $T = 2$. We choose any vector A such that $A \notin T$. Then we have $V = T + \langle A \rangle$. From

Lemma 1.1 we obtain dim $\langle A \rangle^\perp \geq 2$, hence there exists a vector $B \neq O$ in $T \cap \langle A \rangle^\perp$. By Lemma 1.2 we have $T \subset \langle B \rangle^\perp$, thus T, $\langle A \rangle \subset \langle B \rangle^\perp$, and therefore $V = T + \langle A \rangle \subset \langle B \rangle^\perp$ and $B \in V^\perp \cap T$. Since $V^\perp \cap T \subset \mathrm{rad}\,(V, Q)$, we obtain $1 \leq \dim V^\perp \cap T \leq \dim \mathrm{rad}\,(V, Q)$ and thus $\mathrm{Ind}\,(V, Q) = \dim T - \dim \mathrm{rad}\,(V, Q) \leq 1$. This proves Lemma 1.7. ∎

In this book we are interested only in the case of a three-dimensional vector space (V, Q) such that $\dim V^\perp \leq 2$. Obviously, we then have $Q \neq 0$ and $\dim \mathrm{rad}\,(V, Q) \leq 2$. The next lemma shows that the case $\dim V^\perp = 3$, and thus $V = V^\perp$, is a very special case (compare also Lemma 1.11).

LEMMA 1.8. *Let* (V, Q) *be any three-dimensional vector space over a field* K *and dim* $V^\perp = 3$. *Then Ind* $(V, Q) = 0$ *and, if* $Q \neq 0$, *then char* K $= 2$.

PROOF. Let (V, Q) be a metric vector space such that $\dim V = 3 = \dim V^\perp$. Then every isotropic vector is in $V^\perp$ and therefore in $\mathrm{rad}\,(V, Q)$. By Lemma 1.6 we have $\mathrm{Ind}\,(V, Q) = 0$.

If $Q \neq 0$, then we choose a nonisotropic vector X and obtain the following from equation 1: $2 = Q(X)^{-1} g_Q(X, X) = 0$, hence char $K = 2$. ∎

Our general assumption $\dim V^\perp \leq 2$ for a three-dimensional vector space (V, Q) implies, if the field K is of char 2, that $\dim V^\perp \leq 1$. Moreover, we wish to prove:

LEMMA 1.9. *Let* (V, Q) *denote a three-dimensional metric vector space over a field of char* 2. *Then dim* $V^\perp \neq 2$.

PROOF. Suppose there is a three-dimensional metric vector space over a field K of char 2 such that $\dim V^\perp = 2$. Then we choose two linear independent vectors A, B in $V^\perp$ and any vector C not in $V^\perp$. Then A, B, C is a basis of V. From $g_Q(aA + bB + cC, C) = a g_Q(A, C) + b g_Q(B, C) + c g_Q(C, C) = 0$ for all a, b, c in K we obtain $C \in V^\perp$, a contradiction. Thus our assumption is false, and Lemma 1.9 is proved. ∎

3. REPRESENTATION IN A BASIS

Let (V, Q) be a finite-dimensional metric vector space and $E_1, \ldots, E_n$ a basis of V. Then for any vector X in V there is a unique representation $X = \sum_{i=1}^n x_i E_i$ such that $x_1, \ldots, x_n \in K$. This implies

$$Q(X) = \sum_{i=1}^n x_i^2 Q(E_i) + \sum_{\substack{i,j=1 \\ i<j}}^n x_i x_j g(E_i, E_j) \tag{3}$$

Thus the quadratic form is uniquely determined by the elements $Q(E_1), \ldots, Q(E_n)$ and $g(E_1, E_2), \ldots, g(E_{n-1}, E_n)$ in K.

Conversely, for any set of elements $a_1, \ldots, a_n$; $b_{12}, b_{23}, \ldots, b_{n-1 n}$ in K

$$Q(X) = \sum_{i=1}^{n} x_i^2 a_i + \sum_{\substack{i,j=1 \\ i<j}}^{n} x_i x_j b_{ij} \quad \text{if} \quad X = \sum_{i=1}^{n} x_i E_i$$

defines a quadratic form over V.

4.   REFLECTIONS AND THE ORTHOGONAL GROUP

Let (V, Q) be a three-dimensional metric vector space and $g := g_Q$ the bilinear form belonging to the quadratic form Q. Then for any nonisotropic vector A, we consider the mapping

$$\sigma : \begin{cases} V \to V \\ X \mapsto -X + \dfrac{g(X, A)}{Q(A)} A \end{cases}$$

Here σ is a linear mapping of V onto itself, and the following properties can easily be computed:

(1) $\sigma^2 = 1$.

(2) $Q(X\sigma) = Q(X)$ *for all* X *in* V.

(3) $det\,\sigma = 1$.

Obviously, (2) implies:

(4) $g(X\sigma, Y\sigma) = g(X, Y)$ *for all* X, Y *in* V.

To indicate the dependence of σ on the vector A, we denote the map σ more precisely by σ_A.

The map σ_A for any nonisotropic vector A is called the *reflection* of the metric vector space (V, Q) in the vector A. The set of all the reflections in nonisotropic vectors is denoted by $S(V, Q)$. We call the group generated by $S(V, Q)$ the *orthogonal group* of the metric vector space (V, Q), and denote it by $O(V, Q)$.* If $S(V, Q) = \varnothing$, we put $O(V, Q) := \{1\}$.

For brevity we introduce the notation

$$X \circ Y := \frac{g(X, Y)}{Q(Y)} \qquad X, Y \in V \quad \text{such that} \quad Q(Y) \neq 0$$

---

* It should be observed that our definition of an orthogonal group coincides with that given in reference 2; however, it differs from other definitions given in the literature.

Then the "product" ∘ is a mapping of $V \times V$ in K and is linear in the left argument, and we have

$$X \circ Y = 0 \Leftrightarrow g(X, Y) = 0$$

Furthermore, we can write

$$X\sigma_A = -X + (X \circ A)A \quad \text{for all} \quad X \in V$$

For further application we collect some little statements on the reflections and the orthogonal group of a metric vector space.

LEMMA 1.10. *Let* (V, Q) *be a three-dimensional metric vector space over the field* K *and* $\alpha, \beta \in O(V, Q)$. *Then* $\alpha = s\beta$ *for* $s \in K$ *implies* $s = 1$.

PROOF. Our assertion follows immediately if $S(V, Q) = \varnothing$. Now let $S(V, Q) \neq \varnothing$. Then there is a nonisotropic vector X. Applying (2) we obtain $s^2 = s^2 Q(X\beta)Q(X)^{-1} = Q(s(X\beta))Q(X)^{-1} = Q(X(s\beta))Q(X)^{-1} = Q(X\alpha)Q(X)^{-1} = 1$. From (3) we obtain $s^3 = s^3 \det \beta = \det(s\beta) = \det \alpha = 1$, hence $s = 1$. ∎

Let (V, Q) denote a metric vector space over a field K and $L(V, Q)$ the set of all one-dimensional nonisotropic subspaces of (V, Q). Then

$$\Phi(V, Q): \begin{cases} L(V, Q) \to S(V, Q) \\ \langle A \rangle \mapsto \sigma_A \end{cases}$$

is a well defined map, since $\langle A \rangle = \langle B \rangle$ implies there is an element $a \neq 0$ in K such that $B = aA$ and by definition of a reflection in a vector we have $\sigma_B = \sigma_{aA} = \sigma_A$. We wish to prove:

LEMMA 1.11. *Let* (V, Q) *be a three-dimensional metric vector space such that* $Q \neq 0$. *Then the following properties are equivalent:*

(a) *dim* $V^\perp \neq 3$.

(b) $\Phi(V, Q)$ *is injective.*

(c) $O(V, Q) \neq \{1\}$.

PROOF. (a) $\Rightarrow$ (b): Let dim $V^\perp \neq 3$ and let $\sigma_A = \sigma_B$; then A, B are nonisotropic, and by definition $(X \circ A)A = (X \circ B)B$ for all X in V. If $A \in V^\perp$, then $(X \circ A) = 0$ for all X in V, hence $B \in V^\perp$. If dim $V^\perp \leq 1$, then $\langle A \rangle = V^\perp = \langle B \rangle$, as required. The case dim $V^\perp \geq 2$, thus dim $V^\perp = 2$, is not possible: From Lemma 1.9 we deduce char $K \neq 2$ and therefore equation 1 implies that A, B would be isotropic [since $g_Q(A, A) = 0 = g_Q(B, B)$ for $A, B \in V^\perp$], a contradiction. Now let A be not in $V^\perp$. Then we have

$B \notin V^{\perp}$, and there is a vector X such that $X \notin \langle B \rangle^{\perp}$. Then $X \circ B \neq 0$, and we can put $a := (X \circ A)(X \circ B)^{-1}$ and obtain $B = aA$, hence $\langle A \rangle = \langle B \rangle$, as required. Thus $\Phi(V, Q)$ is injective.

(b) $\Rightarrow$ (c): Let $\Phi(V, Q)$ be injective. It is sufficient to show that $L(V, Q)$ contains at least two elements, for then the injectivity of Φ shows that $S(V, Q)$ contains at least two reflections, and thus that $O(V, Q) \neq \{1\}$. Clearly, $L(V, Q) \neq \varnothing$, as otherwise $Q = 0$. Now suppose $\langle A \rangle \in L(V, Q)$. By Lemma 1.1, $\dim \langle A \rangle^{\perp} \geq 2$, hence there exists $B \in \langle A \rangle^{\perp}$ such that $\langle A \rangle \neq \langle B \rangle$. If $\langle B \rangle \in L(V, Q)$, we are done. If $Q(B) = 0$, then $Q(A + B) = Q(A) \neq 0$, and thus $\langle A + B \rangle \in L(V, Q)$, and $\langle A + B \rangle \neq \langle A \rangle$. This proves (c).

(c) $\Rightarrow$ (a): Let $O(V, Q) \neq \{1\}$; then there is at least one reflection $\sigma_A \neq 1$ in $S(V, Q)$. If $A \in V^{\perp}$, then equation 1 yields $2 = g_Q(A, A)Q(A)^{-1} = 0$ and thus char $K = 2$. By definition of a reflection we obtain $X\sigma_A = X$ for all X in V and therefore $\sigma_A = 1$, a contradiction. Hence $A \notin V^{\perp}$ and $\dim V^{\perp} \neq 3$. This proves (a) and concludes the proof of Lemma 1.11. ∎

Lemma 1.11 shows that the special case $\dim V^{\perp} = 3$ for a three-dimensional metric vector space (V, Q) is of limited interest, since the orthogonal group $O(V, Q)$, which we may regard as an automorphism group of the metric vector space, is trivial.

2
INCIDENCE
STRUCTURES

We require as general a definition of metric incidence structures as possible. We therefore first introduce the notion of a ternary equivalence relation.

1. TERNARY EQUIVALENCE RELATIONS

Let M be a nonempty set and κ a subset of $M \times M \times M$. Then κ is called a *ternary equivalence relation* if it satisfies the following properties:

(E1) (Reflexivity). *If* a, b, *and* c *are not mutually distinct, then* $(a, b, c) \in \kappa$.

(E2) (Symmetry). *If* $(a, b, c) \in \kappa$ *and* π *is a permutation of* $\{a, b, c\}$, *then* $(\pi(a), \pi(b), \pi(c)) \in \kappa$.

(E3) (Transitivity). $a \neq b$ *and* $(a, b, c) \in \kappa$ *and* $(a, b, d) \in \kappa$ *imply that* $(a, c, d) \in \kappa$.

We see that the relation of "three lines are concurrent" applied to the set of all the lines of a Euclidean plane is an example of such a ternary equivalence relation.

The following lemma can easily be proved:

LEMMA 2.1. *Let* κ *be a ternary equivalence relation in a set* M. *For distinct* a *and* b *in* M, *we define*

$$M(a, b) := \{x \mid (a, b, x) \in \kappa\} \tag{$\circ$}$$

Then $a, b \in M(a, b)$, *and*

(a) $x, y, z \in M(a, b)$ *implies that* $(x, y, z) \in \kappa$.
(b) $x, y \in M(a, b), M(c, d)$ *implies that* $x = y$ *or* $M(a, b) = M(c, d)$.

9

PROOF.   By (E1) we have that (a, b, a), $(a, b, b) \in \kappa$, and therefore $a, b \in M(a, b)$.

(a)   Let $x, y, z \in M(a, b)$. If $x = y$, then $(x, y, z) \in \kappa$ by (E1). So assume that $x \neq y$. Then $a \neq x$ or $a \neq y$. Without loss of generality, we may assume that $a \neq x$. Since (a, b, x), $(a, b, y) \in \kappa$, we can conclude from (E3) that $(a, x, y) \in \kappa$. Analogously, we see that $(a, x, z) \in \kappa$. Hence we can use (E2) to conclude that (x, a, y), $(x, a, z) \in \kappa$. Since $x \neq a$, we deduce from (E3) that $(x, y, z) \in \kappa$, as required.

(b)   Let $x, y \in M(a, b)$, $M(c, d)$ and assume that $x \neq y$. Choose any $z \in M(a, b)$. Then from (a) we know that $z \in M(x, y)$, hence $M(a, b) \subset M(x, y)$. In particular, $a, b \in M(x, y)$. Reversing the roles of a, b and x, y we obtain $M(x, y) \subset M(a, b)$, hence $M(a, b) = M(x, y)$. Similarly, $M(c, d) = M(x, y)$, and therefore $M(a, b) = M(c, d)$, as required. ∎

2. INCIDENCE STRUCTURES

Next we define an incidence structure. Let L be a set containing at least two elements, and let κ be a ternary equivalence relation on L. Then the pair (L, κ) is called an *incidence structure*. Elements of L, denoted by lower case latin letters, are called *lines*; three lines a, b, c are called *concurrent* if $(a, b, c) \in \kappa$. The subsets $L(a, b)$ of L defined by κ according to $(\circ)$ are called *points*; they are henceforth denoted by capital letters. A line a is said to be *incident* with a point B if and only if $a \in B$. Two points A and B are said to be *connected* (by a line) if $A \cap B \neq \varnothing$. From Lemma 2.1 we have

$$a, b \in A, B \quad implies \quad a = b \quad or \quad A = B$$

Three lines a, b, c are concurrent if and only if there exists a point incident with each of a, b, c.

For every point A, there are at least two lines incident with A.

In the following we use these three statements without referring to them explicitly.

The point A is Δ-*connected* (dreiseitverbindbar) if A is connected with at least one point of any triple B, C, D of distinct, pairwise connected points. The point A is 1-Δ-*connected* if it is Δ-connected and if there exists at least one such triple of points B, C, D for which A is connected with exactly one point of the triple. We say that A is 3-Δ-*connected*, or *completely connected*, if A is connected with all points. A Δ-connected point that is neither completely connected nor 1-Δ-connected is called a 2-Δ-*connected point*. Obviously, a 2-Δ-connected point is not connected with all points; it is, however, connected with all but at most one point of any given line.

An incidence structure (L, κ) is called Δ-*connected* if every point of (L, κ) is Δ-connected.

Let (L, κ) and (L', κ') be two incidence structures. Then a bijective mapping α from L to L' is called an *isomorphism* if

$$(a, b, c) \in \kappa \Leftrightarrow (a\alpha, b\alpha, c\alpha) \in \kappa'$$

If $(L, \kappa) = (L', \kappa')$, we call α an *automorphism* or a *collineation*. Since an isomorphism α maps concurrent lines onto concurrent lines, it maps points onto points. Moreover, we have

$$L(a, b)\alpha = L(a\alpha, b\alpha)$$

[This follows from the following equivalent statements: $x' \in L(a, b)\alpha \Leftrightarrow \exists y$ such that $x' = y\alpha$ and $y \in L(a, b) \Leftrightarrow \exists y$ such that $x' = y\alpha$ and $(a, b, y) \in \kappa \Leftrightarrow \exists y$ such that $x' = y\alpha$ and $(a\alpha, b\alpha, y\alpha) \in \kappa' \Leftrightarrow \exists y$ such that $x' = y\alpha$ and $y\alpha \in L(a\alpha, b\alpha) \Leftrightarrow x' \in L(a\alpha, b\alpha)$.]

The collineations of an incidence structure (L, κ) form a group.

Let α be any collineation. A line a is called an *axis* of α if $a\alpha = a$ and if all the points on a are fixed by α, i.e. if $A\alpha = A$ for all points A on a. A point A is called a *center* of α if all the lines in A are fixed by α. A collineation with an axis a is called an *axial collineation*, and a collineation with a center a *central collineation*.

LEMMA 2.2. *If β is a collineation and α a collineation with an axis a (or a center A) of an incidence structure, then $\beta^{-1}\alpha\beta$ is a collineation with the axis $a\beta$ (or with the center $A\beta$).*

PROOF. Let a be an axis of α. Then $a\beta(\beta^{-1}\alpha\beta) = a\beta$ (as $a\alpha = a$). Furthermore, $a\beta \in X$ for a point X implies $a \in X\beta^{-1}$, and therefore $X\beta^{-1}\alpha = X\beta^{-1}$ and $X\beta^{-1}\alpha\beta = X$. Thus $\beta^{-1}\alpha\beta$ fixes $a\beta$ and all the points on $a\beta$. Therefore $\beta^{-1}\alpha\beta$ is a collineation with axis $a\beta$.

The other statement of Lemma 2.2 follows analogously. ∎

Let (L, κ) be an incidence structure and L' be a subset of L, containing at least two elements. Let κ' be the restriction of κ to $L' \times L' \times L'$. Then (L', κ') is an incidence structure. We call it a *substructure* of the incidence structure (L, κ).

If X is a point of an incidence structure, we let $|X|$ denote the cardinality of X; similarly, if x is a line, we let $|x|$ be the cardinality of the set of points incident with x.

A set of four lines a, b, c, d in an incidence structure is called a *quadrilateral* if no three of the lines are concurrent.

If (L, κ) is an incidence structure, we denote by $(*)$ the following property:

$$(L, \kappa) \quad \text{contains a quadrilateral} \tag{*}$$

It is instructive to consider the incidence structures for which (∗) is not valid. We first prove

(1)   *If the incidence structure* (L, κ) *contains two points* A, B *that are not connected, then* (∗) *holds.*

For let $A = L(a, b)$, $B = L(a', b')$. Then $A \cap B = \varnothing$, and in particular, $a', b' \notin L(a, b)$ and $a, b \notin L(a', b')$. Thus the four lines a, b, a', b' satisfy (∗).

Hence we have

(2)   *If* (∗) *does not hold in* (L, κ), *then any point in* (L, κ) *is completely connected.*

Next suppose (∗) does not hold in (L, κ), and let A, B be two points of (L, κ) which are connected by a line c. If A contains two other lines a, $a' \neq c$, and if B contains two other lines b, $b' \neq c$, then again the set a, a', b, b' forms a quadrilateral. Hence either all points consist of precisely two lines, or there exists a unique point A that contains more than two lines. In the former case $L = \{a, b, c\}$ with points $A = L(a, c)$, $B = L(b, c)$, and $C = L(a, b)$. In the latter case there exists exactly one line $z \neq A$. Thus we have proved the following.

Let (L, κ) be an incidence structure in which (∗) does not hold. Then any point in (L, κ) is completely connected, and (L, κ) is one of the following:

(a)   L *is a point or*
(b) $L = L' \cup \{z\}$ *where* L′ *is a point, and* z *is a line not in* L′. (*If* L′ *contains exactly two lines, then* L *is a triangle.*)

For other examples of incidence structures compare Section 3, 4, and 5.

3. PROJECTIVE INCIDENCE STRUCTURES AND SUBSTRUCTURES

First we establish some important examples of incidence structures.

Let (P, L, I) be a projective plane in the sense of the Appendix. Then three lines a, b, c in L are concurrent in the projective plane if there is a point A in P such that $A\,\mathrm{I}\,a, b, c$. If we put

$$\kappa := \{(a, b, c) \mid a, b, c \in L \quad \text{and} \quad a, b, c \quad \text{are concurrent}\}$$

then (L, κ) is an incidence structure. We call it a *projective incidence structure*.

Now we consider some special substructures (L', κ') of a projective incidence structure (L, κ). It suffices that we are given the set $L \setminus L'$.

(a) $L \setminus L'$ *contains exactly one line* z. *Then we call* (L', κ') *an affine incidence structure.*

In an affine incidence structure (L', κ') we can introduce a parallel relation by defining two lines a, b to be *parallel* if and only if they are concurrent with the line z, used in the definition of the affine incidence structure. The parallel relation is an equivalence relation, and the equivalence classes are called *parallel pencils*.

Clearly, one may derive an affine incidence structure as well from the concept of an affine plane in the sense of the Appendix: Let (P', L', I') be any affine plane, and let

$$\kappa' := \{(a, b, c) \mid a, b, c \in L' \quad \text{and} \quad a, b, c \quad \text{are concurrent or} \quad a, b, c \quad \text{are parallel}\}$$

Then (L', κ') is an incidence structure that is obviously isomorphic to an affine incidence structure.

(b) $L \setminus L'$ *consists of a point, thus of the set of lines defining a point in the projective incidence structure* (L, κ). *We call* (L', κ') *a star-complement.*

We saw that an affine incidence structure is obtained from a projective one by the removal of a line. We observe here that a star-complement is obtained by the dual process of removing a point from the given projective plane. Hence we see that a star-complement is precisely a dual affine plane.

(c) $L \setminus L'$ *consists of exactly two distinct points. We may call* (L', κ') *a double star-complement.*

(d) $L \setminus L'$ *is an oval, that is, a set of lines that contains at least three lines such that no three mutually distinct lines are concurrent. We call* (L', κ') *an oval-complement.*

Both the projective incidence structures, as well as all the incidence structures defined under (a), (b), (c), and (d) are Δ-connected.

In particular, every point consisting of a parallel pencil in an affine incidence structure is 2-Δ-connected, whereas all the other points are completely connected. If (L', κ') is a star-complement, then all its points are 2-Δ-connected. For a double star-complement, there are no completely connected points. In this case let A and B be the two points that make up the set $L \setminus L'$; then every point of (L', κ') that comes from a point on the line of L connecting A and B is 2-Δ-connected, while all the other points of (L', κ') are 1-Δ-connected. Finally, suppose (L', κ') is an oval-complement. Then all those points of (L', κ') which are derived from points of (L, κ) lying on two distinct lines of the oval are 1-Δ-connected. The remaining points may be either 2- or 3-Δ-connected—either case can occur in specially constructed examples.

A projective incidence structure is called a *Pappian* or a *Desarguesian* projective incidence structure if the theorem of Pappus or the theorem of Desargues, respectively, is valid in the corresponding projective plane (compare the Appendix).

Analogously, we may define a Pappian affine incidence structure, a Desarguesian affine incidence structure, and so on.

4. PROJECTIVE AND AFFINE INCIDENCE STRUCTURES OVER A FIELD

Let K be any field of arbitrary characteristic. To define the affine incidence structure over K, let V be the two-dimensional vector space over K. In the notation of Chapter 1 we denote by $A + \langle B \rangle$ a residue class (coset) of the subspace $\langle B \rangle$, represented by the vector A. Let O be the zero vector in V. We put

$$L := \{ A + \langle B \rangle \mid A, B \in V \quad \text{and} \quad B \neq O \}$$

Then it is well known that the triplet $(V, L, \in)$ is an affine plane (compare the Appendix). The affine incidence structure associated to this affine plane is called the *affine coordinate incidence structure* over the field K, denoted by $AI(V)$ or by $AI(K)$.

Now we define the projective incidence structure over K: Let V be the three-dimensional vector space over K. In the notations of Chapter 1 we put

$$L(V) := \{ \langle A \rangle \mid A \neq O \}$$

$$\kappa(V) := \{ (\langle A \rangle, \langle B \rangle, \langle C \rangle) \mid A, B, C \quad \text{linearly dependent} \}$$

Then $(L(V), \kappa(V))$ is a projective incidence structure. We call it the *projective coordinate incidence structure* over the vector space V or over the field K, and denote it by $I(V)$ or by $I(K)$.

Obviously, any Pappian projective incidence structure is isomorphic to a projective coordinate incidence structure $(L(V), \kappa(V))$ (compare the Appendix).

5. INCIDENCE STRUCTURES OVER A METRIC VECTOR SPACE

Let (V, Q) be a three-dimensional metric vector space over the field K and

$L(V, Q) := \{\langle A \rangle \mid Q(A) \neq 0\}$

$\kappa(V, Q) := \{(\langle A \rangle, \langle B \rangle, \langle C \rangle) \mid Q(A), Q(B), Q(C) \neq 0$

$$\text{and} \quad A, B, C \quad \text{linearly dependent}\}$$

If $I(V) = (L(V), \kappa(V))$ is the projective incidence structure over the field K, then $L(V, Q)$ consists of all nonisotropic lines of $I(V)$, and $\kappa(V, Q)$ is the restriction of $\kappa(V)$ to $L(V, Q) \times L(V, Q) \times L(V, Q)$. We show:

LEMMA 2.3. *Let* (V, Q) *be a three-dimensional metric vector space and* $Q \neq 0$. *Then* $(L(V, Q), \kappa(V, Q))$ *is a* Δ-*connected incidence structure.*

PROOF. Obviously, $\kappa(V, Q)$ is a ternary equivalence relation in $L(V, Q)$. Furthermore, we prove that $L(V, Q)$ contains at least two elements. Thus we must show that (V, Q) contains at least two distinct nonisotropic one-dimensional subspaces $\langle A \rangle$ and $\langle B \rangle$. From $Q \neq 0$ we deduce there exists at least one A such that $Q(A) \neq 0$. By Lemma 1.1 we know that $\dim \langle A \rangle^\perp \geq 2$. Hence we can find some $B \in \langle A \rangle^\perp$ such that $\langle B \rangle \neq \langle A \rangle$. If $Q(B) \neq 0$, then $\langle B \rangle$ is the second nonisotropic one-dimensional subspace required.

If $Q(B) = 0$, then $Q(A + B) = Q(A) + Q(B) + g(A, B) = Q(A) \neq 0$, hence $\langle A + B \rangle$ is the second nonisotropic one-dimensional subspace. Hence $L(V, Q)$ contains at least two elements, and therefore $(L(V, Q), \kappa(V, Q))$ is an incidence structure.

Incidence structure $(L(V, Q), \kappa(V, Q))$ is Δ-connected: By Lemma 1.4, any point in $I(V)$ which contains at least one nonisotropic line contains at most two isotropic lines. Since two points in $(L(V, Q), \kappa(V, Q))$ are not connected by a line in $L(V, Q)$ if and only if the connecting line of the points in $I(V)$ is isotropic, Lemma 1.4 implies that every point in $(L(V, Q), \kappa(V, Q))$ is Δ-connected, hence $(L(V, Q), \kappa(V, Q))$ is Δ-connected. This proves the lemma. ■

If (V, Q) is a three-dimensional vector space such that $Q \neq 0$, then we call $(L(V, Q), \kappa(V, Q))$ the *incidence structure over the metric vector space* (V, Q) and denote it by $I(V, Q)$. Clearly, $I(V, Q)$ is a substructure of $I(V)$.

More detailed information on the incidence structure $I(V, Q)$ can be given if we discuss some special cases:

PROPOSITION 2.4. *Let* (V, Q) *be a three-dimensional metric vector space and* $Q \neq 0$. *Then one of the following cases occurs:*

(a) *Ind* $(V, Q) = 0$ *and dim rad* $(V, Q) = 0$ *and* $I(V, Q)$ *is the projective incidence structure* $I(V)$.

(b) *Ind* $(V, Q) = 0$ *and dim rad* $(V, Q) = 1$ *and* $I(V, Q)$ *is an affine incidence structure.*

(c) *Ind* $(V, Q) = 0$ *and dim rad* $(V, Q) = 2$ *and* $I(V, Q)$ *is a star-complement.*

(d) *Ind* $(V, Q) = 1$ *and dim rad* $(V, Q) = 0$ *and* $I(V, Q)$ *is an oval-complement.*

(e) *Ind* $(V, Q) = 1$ *and dim rad* $(V, Q) = 1$ *and* $I(V, Q)$ *is a double star-complement.*

PROOF. Let (V, Q) be a three-dimensional metric vector space and $Q \neq 0$. Then by Lemma 1.7 the only combinations of Ind (V, Q) and dim rad (V, Q) are enumerated in the cases (a) to (e).

(a) Ind $(V, Q) = 0$ and dim rad $(V, Q) = 0$. In this case any vector $\neq O$ is nonisotropic, hence we have $L(V, Q) = L(V)$ and $\kappa(V, Q) = \kappa(V)$, thus $I(V, Q) = I(V)$.

(b) Ind $(V, Q) = 0$ and dim rad $(V, Q) = 1$. By Lemma 1.6, rad (V, Q) is the only isotropic one-dimensional subspace of V, hence $L(V) \setminus L(V, Q)$ consists of exactly one line, and $I(V, Q)$ is an affine incidence structure.

(c) Ind $(V, Q) = 0$ and dim rad $(V, Q) = 2$. By Lemma 1.6 all the isotropic one-dimensional subspaces are contained in the two-dimensional isotropic subspace rad (V, Q), hence $I(V, Q)$ is a star-complement.

(d) Ind $(V, Q) = 1$ and dim rad $(V, Q) = 0$. Then any two-dimensional subspace is nonisotropic. Furthermore, the set of all isotropic one-dimensional subspaces is an oval: We first show there are at least three distinct isotropic lines. From Ind $(V, Q) = 1$ we deduce there is at least one isotropic line $\langle A \rangle$. Since dim rad $(V, Q) = 0$, we have $A \notin V^{\perp}$, hence by Lemma 1.1, dim $\langle A \rangle^{\perp} = 2$ and there is a line $\langle B' \rangle$ such that $B' \notin \langle A \rangle^{\perp}$, $\langle A \rangle$. Then the vector $B := (Q(B')g(A, B')^{-1})A - B'$ is isotropic and $\langle B \rangle \neq \langle A \rangle$. There is a line $\langle C' \rangle$ such that $C' \notin \langle A \rangle^{\perp}$, $\langle A, B \rangle$ and $C := (Q(C')g(A, C')^{-1})A - C'$ is isotropic and $\langle C \rangle \neq \langle A \rangle$, $\langle B \rangle$. Hence there are three distinct isotropic lines. By Lemma 1.4 no three distinct isotropic lines are concurrent, hence the set of isotropic lines is an oval, and $I(V, Q)$ is an oval-complement.

(e)   Ind $(V, Q) = 1$ and dim rad $(V, Q) = 1$. By definition of Ind (V, Q) there is a two-dimensional isotropic subspace T. By Lemma 1.5, rad $(V, Q) \subset T$. Let A be any vector in $T \setminus$ rad (V, Q); then $A \notin V^\perp$ and, by Lemma 1.1, dim $\langle A \rangle^\perp = 2$. Moreover, by Lemma 1.2, $T \subset \langle A \rangle^\perp$, hence $T = \langle A \rangle^\perp$. Choose any nonisotropic vector B. Then $B \notin \langle A \rangle^\perp$ and $C := (Q(B)g(A, B)^{-1})A - B$ is isotropic and $C \notin T$. Put $T' := \langle C \rangle +$ rad (V, Q). Then $T' \neq T$ and dim $T' = 2$ and, by Corollary 1.3, T' is isotropic. We show that every isotropic vector X is in $T \cup T'$: If we assume this is not true, then $T'' = \langle X \rangle +$ rad (V, Q) is an isotropic two-dimensional subspace $\neq T, T'$ (compare Corollary 1.3). Let U be any two-dimensional subspace such that rad (V, Q) is not in U. Then there are three distinct lines in U and in T, T', T'', respectively. These lines are isotropic and therefore U is isotropic (compare Lemma 1.4). From Corollary 1.3 we deduce that $V = U +$ rad (V, Q) is isotropic, in contradiction to $Q \neq 0$. Hence $X \in T \cup T'$, and $I(V, Q)$ is a double star-complement.

This completes the proof of Proposition 2.4.   ■

All the cases mentioned in Proposition 2.4 occur: Let $V = \mathbf{R}^3$ and Q_1, Q_2, Q_3, Q_4, Q_5 be the quadratic forms on V defined by [we write $Q(x_1, x_2, x_3)$ instead of $Q((x_1, x_2, x_3))$]:

$$\left. \begin{array}{l} Q_1(x_1, x_2, x_3) = x_1{}^2 + x_2{}^2 + x_3{}^2 \\ Q_2(x_1, x_2, x_3) = x_1{}^2 + x_2{}^2 \\ Q_3(x_1, x_2, x_3) = x_1{}^2 \\ Q_4(x_1, x_2, x_3) = x_1{}^2 + x_2{}^2 - x_3{}^2 \\ Q_5(x_1, x_2, x_3) = x_1{}^2 - x_2{}^2 \end{array} \right\} \forall (x_1, x_2, x_3) \in \mathbf{R}^3$$

Then Ind $(V, Q_1) = 0$ and   dim rad $(V, Q_1) = 0$;   Ind $(V, Q_2) = 0$ and dim rad $(V, Q_2) = 1$; Ind $(V, Q_3) = 0$ and dim rad $(V, Q_3) = 2$; Ind $(V, Q_4) = 1$ and dim rad $(V, Q_4) = 0$; Ind $(V, Q_5) = 1$ and dim rad $(V, Q_5) = 1$.

LEMMA 2.5. *Let (V, Q) be a three-dimensional vector space over the field K and $Q \neq 0$. If $K = GF(2)$, let Ind $(V, Q) = 0$. Then the incidence structure $I(V, Q)$ contains a quadrilateral.*

PROOF. This follows immediately from Proposition 2.4 since, if $K \neq GF(2)$, then every point of $L(V)$ contains at least four lines and, if $K = GF(2)$, then we obtain the assertion from (a) to (c) in Proposition 2.4.   ■

3
METRIC CONCEPTS

In general, the theory of metrics is related to measuring a segment on a line; more precisely, a metric is a map of the set of pairs of points into the field of reals, satisfying certain conditions.* This concept does not work in a purely synthetic development of metric geometry, even if we replace the field of the reals by any other field. Thus we have to look for a purely synthetic definition of metrics. At the present stage there is no general agreement about the best introduction of a synthetic definition of a metric. In this book we propose two different definitions. We introduce the concept of orthogonality and—independently—the concept of a reflection. Each of these notions is added to the concept of an incidence structure to define metric structures. Adding the orthogonality, we arrive at the concept of an incidence structure with orthogonality and, adding reflections, we obtain the concept of an incidence structure with reflections. The latter notion is related to one of the main topics of this book, the theory of S-groups and their group planes.

One of the more important tasks in our investigations is to look for conditions such that the given incidence structure, with orthogonality or with reflections, can be embedded in the projective incidence structure over a vector space V such that the given metrics can be represented by a quadratic form Q on V.

To familiarize the reader with some aspects of metric planes and to give some examples, we begin with a brief and elementary exposition of classical Euclidean and non-Euclidean planes.

1. CLASSICAL EUCLIDEAN AND NON-EUCLIDEAN PLANES

(a) Let K be any field of char $\neq 2$ and let $V = K \times K = K^2$ be the two-dimensional vector space over K. Let $AI(V)$ be the affine incidence

---

* Compare the theory of metric spaces in the frame of general topology.

structure over the field K. Finally, let k be any element of K such that $-k$ is not a square. Then for $A := (a, a')$ and $B := (b, b')$ in V we put

$$A \circ B := ab + ka'b' \tag{1}$$

We may introduce the notion of orthogonality in $AI(V)$ as follows: Two lines $A + \langle B \rangle$ and $A' + \langle B' \rangle$ are said to be *orthogonal* if $B \circ B' = 0$. Then we may speak of a *Euclidean plane E* over K. To express the dependency of E on the field K and the element k, we may write more explicitly $E(K, k)$. The set of all perpendiculars to a line is a point. For every line a there is exactly one involutory collineation of $AI(V)$ with axis a which fixes every perpendicular to a. This collineation may be called the *reflection* in the line a. Let $S(E)$ be the set of all reflections in lines of E, and let $B(E)$ be the group generated by $S(E)$. Then $B(E)$ is called the *group of motions* of the Euclidean plane E over K.

If K is the field of reals and $k = 1$, the orthogonality in E is the familiar orthogonality in the real Euclidean plane.

(b) Let K be a field of char $\neq 2$ and $V := K^2$, and let $AI(V) = (L, \kappa)$ be the affine incidence structure over K. Moreover, let k be a square $\neq 0$ in K; then we replace equation 1 by the following definition of a product:

$$A \cdot B = ab - ka'b' \tag{2}$$

Let

$$L' := \{A + \langle B \rangle \mid B \cdot B \neq 0\}$$

and let κ' be the restriction of κ to $L' \times L' \times L'$. Then (L', κ') is a substructure of the affine incidence structure $AI(V)$ over the field K. Looking at the projective extension of the affine incidence structure $AI(V)$, we see that (L', κ') is isomorphic to a double star-complement. Two lines $A + \langle B \rangle$ and $A' + \langle B' \rangle$ in L' are said to be orthogonal if $B \cdot B' = 0$. Then we may speak of the *Minkowskian plane M* over the field K.

As in the Euclidean case the set of perpendiculars to a line is a point, and for every line there exists one and only one reflection in that line if we define a reflection in the line a as in the Euclidean case. By $S(M)$ we denote the set of reflections in lines, and by $B(M)$ the group generated by the set $S(M)$.

If K is the field of reals and if we write the vectors of V as (x, t) and put $k = c^2$, where c is the velocity of light, then we may interpret V as the space-time continuum of special relativity and the Lorentz transformations

$$x' := \frac{x - vt}{\sqrt{1 - v^2/c^2}} \qquad t' := \frac{t - (v/c^2)x}{\sqrt{1 - v^2/c^2}}$$

are elements of $B(M)$. We remark that the Lorentz transformations, together with the translations of the plane and the reflections in two orthogonal lines, generate the group $B(M)$.

Now we wish to introduce the real elliptic plane.

(c) Let $V := \mathbf{R}^3$ be the three-dimensional vector space over the reals. Let XY denote the usual inner product, as given by

$$XY = x_1 y_1 + x_2 y_2 + x_3 y_3$$

if $X = (x_1, x_2, x_3)$ and $Y = (y_1, y_2, y_3)$ in V. Moreover, let $I(V) = (L, \kappa)$ be the projective incidence structure over $\mathbf{R}$. Then we define a relation ω of orthogonality by

$$\omega := \{(\langle A \rangle, \langle B \rangle) \mid A, B \neq O \quad \text{and} \quad AB = 0\}$$

and arrive at the *real elliptic plane*.

By definition a line of the real elliptic plane is a line of the three-dimensional real affine space $\mathbf{R}^3$, which goes through the origin, and three lines of the real elliptic plane are concurrent if and only if they are coplanar in the space $\mathbf{R}^3$. If we let P denote the set of points of the real elliptic plane, then $(P, L, \in)$ is a projective plane isomorphic to the real projective plane. Two lines of the real elliptic plane are orthogonal if and only if they are orthogonal as lines of the space $\mathbf{R}^3$. It can easily be proved that, given any line of the real elliptic plane, there exists a reflection in that line. This reflection is defined as an involutory homology with axis a fixing the perpendiculars to a, and is induced by the plane reflection of $\mathbf{R}^3$ in the plane through the origin which is orthogonal to the given line.

We obtain another interpretation of the real elliptic plane, dual to the original one, by intersecting the lines and planes of the space $\mathbf{R}^3$ which pass through the origin with the unit sphere. "Points" are the pairs of opposite points on the sphere (such as north pole and south pole), and three points are collinear if and only if they are incident with a great circle. Thus the "lines" correspond to the great circles, and two lines are orthogonal if and only if the planes containing the great circles are orthogonal in the space $\mathbf{R}^3$.

(d)   Klein's model of the real hyperbolic plane. Without going into the minute details, Klein's model of the real hyperbolic plane may be described as follows: Let k be the unit circle in the real plane $\mathbf{R}^2$, and let L be the set of lines of the real plane which are incident with at least one point of the interior of k. Furthermore, let κ' be the restriction of the relation κ of the affine incidence structure $AI(V)$ defined in (a) to $L \times L \times L$ and let ω be the set of pairs $(a, b) \in L \times L$ such that a is incident with the pole of b in the projective closure of $\mathbf{R}^2$ with respect to the polarity corresponding to k. This yields Klein's model of the real hyperbolic plane. We can prove that for any line a in L there is a reflection in the line a, that is, an involutory collineation with axis a fixing all the perpendiculars to the line a. Moreover, the set of perpendiculars to a is a point [in the sense of the incidence structure (L, κ')].

2. INCIDENCE STRUCTURES WITH ORTHOGONALITY

Let (L, κ) be an incidence structure, and let ω be an irreflexive, symmetric binary relation on L; thus $\omega \subset L \times L$. Then the triplet (L, κ, ω) is called an *incidence structure with orthogonality*, and two lines a, b are said to be *orthogonal* if $(a, b) \in \omega$. This relation is also denoted $a \perp b$. If g is a line, we define

$$g^\perp := \{x \mid x \perp g\}$$

We call $g^\perp$ the *set of perpendiculars* of g. A point A is called a *polar point* of g if $A \cap g^\perp$ contains at least two elements. A point A is said to be connected with the set $g^\perp$ of perpendiculars of g if $A \cap g^\perp \neq \varnothing$.

Let (L, κ, ω) and (L', κ', ω') be two incidence structures with orthogonality. Then an isomorphism α of the incidence structure (L, κ) onto the incidence structure (L', κ') is called an *isomorphism* of (L, κ, ω) onto (L', κ', ω') if

$$(a, b) \in \omega \Leftrightarrow (a\alpha, b\alpha) \in \omega'$$

If $(L, \kappa, \omega) = (L', \kappa', \omega')$, we call α an *automorphism* or an *orthogonal collineation*.

Obviously, if α is an orthogonal collineation, then α^{-1} is an orthogonal collineation.

An orthogonal collineation σ with axis g of an incidence structure with orthogonality (L, κ, ω) is called a *reflection in the line* g, sometimes denoted by σ_g, if the following conditions are satisfied:

$$\sigma^2 = 1$$

$$a\sigma = a \quad \text{and} \quad a \neq g \quad \text{if and only if} \quad (a, g) \in \omega$$

By definition, σ fixes all the points on g and all the lines in $g^\perp$.

LEMMA 3.1. *If α is an orthogonal collineation of an incidence structure with orthogonality and σ_g is a reflection in the line* g, *then $\alpha^{-1}\sigma_g\alpha$ is a reflection in the line* $g\alpha$.

PROOF. Clearly, $\alpha^{-1}\sigma_g\alpha$ is a collineation, and we obtain $(a, b) \in \omega \Leftrightarrow$ $(a\alpha^{-1}, b\alpha^{-1}) \in \omega \Leftrightarrow (a\alpha^{-1}\sigma_g, b\alpha^{-1}\sigma_g) \in \omega \Leftrightarrow (a\alpha^{-1}\sigma_g\alpha, a\alpha^{-1}\sigma_g\alpha) \in \omega$. By

Lemma 2.2, $\alpha^{-1}\sigma_g\alpha$ is an axial collineation with axis $g\alpha$. Furthermore, $(\alpha^{-1}\sigma_g\alpha)^2 = \alpha^{-1}\sigma_g\alpha\alpha^{-1}\sigma_g\alpha = 1$, and we have the following equivalent statements: $a\alpha^{-1}\sigma_g\alpha = a$ and $a \neq g\alpha \Leftrightarrow (a\alpha^{-1})\sigma_g = a\alpha^{-1}$ and $a\alpha^{-1} \neq g \Leftrightarrow (a\alpha^{-1}, g) \in \omega \Leftrightarrow (a, g\alpha) \in \omega$. Thus $\alpha^{-1}\sigma_g\alpha$ is a reflection in the line $g\alpha$. ∎

The Euclidean and Minkowskian planes over an arbitrary field K of char $\neq 2$, as well as the real elliptic plane, and Klein's model of the real hyperbolic plane discussed in the preceding section, are examples of incidence structures with orthogonality. Further examples of incidence structures with orthogonality, especially those over fields of char 2, can be deduced from Section 3.

3. METRIC INCIDENCE STRUCTURE OVER A METRIC VECTOR SPACE

Let (V, Q) be any three-dimensional metric vector space and $Q \neq 0$. In the incidence structure $I(V, Q)$ we can introduce an orthogonality relation ω by putting

$$\omega(V, Q) := \{(\langle A \rangle, \langle B \rangle) \mid \langle A \rangle \neq \langle B \rangle \quad \text{in} \quad L(V, Q) \quad \text{and} \quad g_Q(A, B) = 0\}$$

We remark that in the case of char $\neq 2$ the condition $\langle A \rangle \neq \langle B \rangle$ is superfluous, since $g_Q(A, B) = 0$ and $Q(A), Q(B) \neq 0$ imply $\langle A \rangle \neq \langle B \rangle$, by Chapter 1, equation 1.

Then $\omega(V, Q)$ is an irreflexive, symmetric binary relation in $L(V, Q)$, and $(L(V, Q), \kappa(V, Q), \omega(V, Q))$ is an incidence structure with orthogonality. It is called the *metric incidence structure over the metric vector space* (V, Q), and is denoted by $MI(V, Q)$.

Up to isomorphisms, our examples of incidence structures with orthogonality (the Euclidean plane, the Minkowskian plane, etc.) were special cases of the metric incidence structures over a metric vector space (V, Q). We elaborate on this in the case of Euclidean planes:

Let $E := E(K, k)$ be a Euclidean plane as described in Section 1a. Let L be the set of lines of E, and let κ be the concurrency relation in E. Moreover, we denote by ω the orthogonality relation in E. We recall that we are given the product in K^2:

$$(x, y) \circ (x', y') = xx' + kyy'$$

Now we consider the vector space (V, K) such that $V := K^3$ and the quadratic form

$$Q : \begin{cases} V \to K \\ (x, y, z) \mapsto x^2 + ky^2 \end{cases} \tag{3}$$

Since $-k$ is not a square in K, $Q((x, y, z)) = 0$ if and only if $x = y = 0$. Hence $(L(V, Q), \kappa(V, Q), \omega(V, Q))$ is defined. We obtain for the symmetric bilinear form g_Q belonging to Q:

$$\tfrac{1}{2}g_Q((x, y, z), (x', y', z')) = xx' + kyy' = (x, y) \circ (x', y')$$

Clearly, $V^\perp = \langle (0, 0, 1) \rangle$ and thus $\dim V^\perp = 1$. We now construct an isomorphism Φ of (L, κ, ω) onto $(L(V, Q), \kappa(V, Q), \omega(V, Q))$ by defining

$$\Phi : \begin{cases} L \to L(V, Q) \\ (a_1, a_2) + \langle (b_1, b_2) \rangle \mapsto \langle (b_1, b_2, a_1 b_2 - a_2 b_1) \rangle \end{cases}$$

We see that Φ is a mapping from L into $L(V, Q)$, since $((a_1, a_2) + \langle (b_1, b_2) \rangle)\Phi$ is independent of the choice of (b_1, b_2) in the one-dimensional subspace $\langle (b_1, b_2) \rangle$, and since $Q(b_1, b_2, a_1 b_2 - a_2 b_1) \neq 0$ (b_1, b_2 are not both zero). It can easily be proved that Φ is an isomorphism of (L, κ, ω) onto $(L(V, Q), \kappa(V, Q), \omega(V, Q))$. Obviously, $\mathrm{Ind}\,(V, Q) = 0$ and $\dim \mathrm{rad}\,(V, Q) = 1$.

We remark that in the Minkowskian case we only have to interchange $x^2 + ky^2$ in equation 3 by $x^2 - ky^2$ and to interpret k as a square $\neq 0$ in K instead of an element in K, such that $-k$ is not a square.

The interpretation of a Euclidean plane (respectively a Minkowskian plane) as a metric incidence structure over a metric vector space enables us to extend the definition of a Euclidean plane (respectively, Minkowskian plane) to the case of char 2:

Let K be any field of arbitrary characteristic containing an element k such that

$$x^2 + y^2 + kxy \neq 0 \quad \text{for all} \quad (x, y) \neq (0, 0) \quad \text{in} \quad K^2$$

and let $V = K^3$. We define

$$Q : \begin{cases} V \to K \\ (x, y, z) \mapsto x^2 + y^2 + kxy \end{cases} \tag{4}$$

Then (V, Q) is a metric vector space such that $V^\perp = \langle (0, 0, 1) \rangle$, and $(L(V, Q), \kappa(V, Q), \omega(V, Q))$ is defined. We are justified in calling $(L(V, Q), \kappa(V, Q), \omega(V, Q))$ a Euclidean plane, since one can easily verify in the case of characteristic $\neq 2$ that $(L(V, Q), \kappa(V, Q), \omega(V, Q))$ is isomorphic to $(L(V, Q'), \kappa(V, Q'), \omega(V, Q'))$ where Q' is the quadratic form shown in equation 3 with $k' := (2 - k)/(2 + k)$ as the element k in that equation.

Note that our definition generalizes that of a Euclidean plane to include char 2.

Analogously we may extend the definition of a Minkowskian plane to the case of char 2.

For a general definition of a Euclidean, a Minkowskian, an elliptic, and a hyperbolic-metric coordinate plane by means of a metric vector space (over a field) compare Section 7.

4. INCIDENCE STRUCTURES WITH REFLECTIONS

Let (L, κ) be an incidence structure and Φ a mapping of L into the set of axial collineations of (L, κ) such that $a\Phi$ is a collineation with axis a and $(a\Phi)^2$ is the identity. Then we call the collineation $a\Phi$ of any line a in L a *reflection in the line a*, and denote it by σ_a. Moreover, the triplet (L, κ, Φ) is called an *incidence structure with reflections*.

Let (L, κ, Φ) and (L', κ', Φ') denote two incidence structures with reflections. We then call an isomorphism α of (L, κ) onto (L', κ') an *isomorphism* of the incidence structure with reflections (L, κ, Φ) onto the incidence structure with reflections (L', κ', Φ') if for all a in L

$$\alpha^{-1}\sigma_a\alpha = \sigma_{a\alpha}$$

with $\sigma_a := a\Phi$ and $\sigma_{a\alpha} := (a\alpha)\Phi'$. If α is an isomorphism of (L, κ, Φ) onto (L', κ', Φ'), then $\alpha^{-1}(\text{Im } \Phi)\alpha = \text{Im } \Phi'$.

We observe that the concept of reflection is also meaningful for an incidence structure with orthogonality; however, this concept is dependent on the orthogonality, whereas there exist incidence structures with reflections in which no meaningful orthogonality can be defined.

More especially, we define the concept of an S-plane as follows: An incidence structure with reflections (L, κ, Φ) is called an *S-plane* if the following properties are satisfied (we put $\sigma_a := a\Phi$ for a in L and $S := \text{Im } \Phi$):

$$\sigma_a\sigma_b\sigma_c \in S \Leftrightarrow (a, b, c)\in \kappa \qquad \text{[S]}$$

$$\text{If} \quad \sigma_a\sigma_b\sigma_c \quad \text{is involutory, then} \quad \sigma_a\sigma_b\sigma_c \in S \qquad \text{[S']}$$

Let (L, κ, Φ) be any S-plane and (L', κ') be an incidence substructure of (L, κ). Then the triplet $(L', \kappa', \Phi_{|L'})$ is called an *S-subplane* of the S-plane (L, κ, Φ) if $\sigma(L')\subset L'$ for all σ in $L'\Phi$.

In this book we are interested only in a special case of the concept of an S-plane, namely in the concept of a complete metric plane, as defined in Section 5. We use the more general concept of an S-plane only to prove Theorem 4.9 in natural generality.

5. COMPLETE METRIC PLANES

Let (L, κ) be a Δ-connected incidence structure, which contains a quadrilateral, and let Φ be a map of L into the set of axial collineations of

(L, κ) such that for $a \in L$, $a\Phi$ is a collineation with axis a. Then the triplet (L, κ, Φ) is called a *complete metric plane* (or a Δ-*connected S-plane*) if for $\sigma_a := a\Phi$ and $S := \text{Im} \, \Phi$

$$\sigma_a\sigma_b\sigma_c \in S \Leftrightarrow (a, b, c) \in \kappa \qquad \qquad [\text{S}]$$

The following lemma shows that any complete metric plane is an S-plane. Thus we may also call the collineation σ_a the *reflection in the line a*. Moreover, isomorphisms of complete metric planes are defined.

LEMMA 3.2. *Let* (L, κ, Φ) *be any complete metric plane and* $\sigma_a := a\Phi$ *for* $a \in L$ *and* $S := \text{Im} \, \Phi$. *Then we have:*

(a) σ_a *has exactly one axis, if* $\sigma_a \neq 1$.

(b) $\sigma_a^2 = 1$ *for all* a *in* L.

(c) Φ *is injective.*

(d) $\sigma_a\sigma_b\sigma_a = \sigma_{b\sigma_a}$ *for all* a, b *in* L.

(e) $\sigma_a\sigma_b = \sigma_b\sigma_a \Leftrightarrow b\sigma_a = b$ *for all* a, b $\in$ L.

(f) *If* $\sigma_a\sigma_b\sigma_c$ *is involutory, then* $\sigma_a\sigma_b\sigma_c \in S$.

PROOF. If X is a set of lines, we let $|X|$ denote the cardinality of X; similarly, if x is a line, we let $|x|$ be the cardinality of the set of points incident with x. We first prove

(1) *If* $\alpha_a\sigma_b = 1$, *then* a $=$ b.

If we assume $\sigma_a\sigma_b = 1$ then $\sigma_a\sigma_b\sigma_x = \sigma_x \in S$ for all $x \in L$. Hence, by [S], we have $(a, b, x) \in \kappa$ for all lines x. If $a \neq b$, this implies that $x \in L(a, b)$ for all lines x, contradicting the existence of a quadrilateral.

(a): Suppose there exists some $a \in L$ such that $\alpha_a \neq 1$, and σ_a has an axis $b \neq a$. Let $A := L(a, b)$. We prove

(2) $x\sigma_a = x$ *for all* $x \in L$ *such that* $x \notin A$.

Choose $x \in L$ but $x \notin A$. Then $x \neq a, b$ since $a, b \in A$. Since a, b are both axes of σ_a, we know that $L(a, x)$, $L(b, x)$ are fixed under σ_a. Hence $x, x\sigma_a \in L(a, x)$, $L(b, x)$, implying that $x\sigma_a = x$ as claimed, or $L(a, x) = L(b, x)$, which yields the contradiction $x \in L(a, b) = A$.

Since $\sigma_a \neq 1$, there exists a line c such that $c\sigma_a \neq c$. By (2), we know that $c \in A$. Since $A\sigma_a = A$, this implies $c\sigma_a \in A$. We now claim that $|X| = 2$ for all points $X \neq A$, with $c \in X$. To prove it suppose there existed a point $X \neq A$ with $c \in X$, and suppose there existed distinct lines $x, y \in X$, and $x, y \notin A$. By (2), we could then deduce that $X\sigma_a = L(x, y)\sigma_a = L(x\sigma_a, y\sigma_a) = L(x, y) = X$. But then $c, c\sigma_a \in A$, X, and $A \neq X$ yields the contradiction $c\sigma_a = c$.* Now suppose $x \notin A$. Then by the above $|L(x, c)| =$

---

* Observe that our proof of (a) would be complete here if we knew—as we do in the case of projective and affine incidence structures—that every point contains at least three lines.

2. Hence $x\sigma_c = x$ and $c\sigma_x = c$. Similarly, we have $|X| = 2$ for all points $X \neq A$ with X on $c\sigma_a$, and $x\sigma_{c\sigma_a} = x$, and $(c\sigma_a)\sigma_x = c\sigma_a$ for all $x \notin A$. Hence A is linewise fixed under all σ_x with $x \notin A$. By assumption there exist lines d, e such that no three of the lines a, b, d, e are concurrent. Let $B := L(c, d)$, and $Z := L(d, e)$. Since B is Δ-connected, we have $|y| \leq |B| + 2$ for all lines $y \notin B$. Since $|B| = 2$, we can thus write $|y| \leq 4$ for all $y \notin B$. We obtain the same inequality if we replace the point B by $B\sigma_a$ and $L(c, e)$; this yields $|x| \leq 4$ for all lines x.

In particular, we have $|d| \leq 4$. The points $L(a, d)$, $L(b, d)$, $L(c, d)$, $L(c\sigma_a, d)$ are mutually distinct, and therefore represent all the points on d. Hence the point Z is equal to one of these four points, and is connnected with A by a line. This implies $|Z| \geq 3$, since $d, e \in Z$ and $d, e \notin A$. Thus $Z = L(a, d)$ or $Z = L(b, d)$. If $Z = L(a, d)$, then [S] yields that there exists a line f such that $\sigma_a\sigma_d\sigma_e = \sigma_f$. From (1) we deduce $a \neq f$ (otherwise we would have $\sigma_d\sigma_e = 1$ and therefore $d = e$). Thus $f \notin A$ and σ_f, σ_d, σ_e fix all the lines in A. Hence $\sigma_a = \sigma_f\sigma_e^{-1}\sigma_d^{-1}$ fixes all the lines in A, a contradiction. Thus we are left with the case $Z = L(b, d)$. The argument just given yields that σ_b fixes all the lines in A. Hence $x\sigma_a\sigma_b\sigma_a = x\sigma_b$ for all x by (2), and therefore $\sigma_a\sigma_b\sigma_a = \sigma_b$. From [S] we deduce that there exists a line r such that $\sigma_a\sigma_b\sigma_c = \sigma_r$. Then $r \neq a, b, c$ by (1) and because $\sigma_a\sigma_b\sigma_a = \sigma_b$. We compute that $(c\sigma_a)\sigma_r = (c\sigma_a)\sigma_a\sigma_b\sigma_c = c \neq c\sigma_a$ and therefore $r \in A$. However, this implies that $r = c\sigma_a$, in contradiction to $(c\sigma_a)\sigma_r = c \neq c\sigma_a$. This concludes the proof of (a).

(b): Choose any line $a \in L$. Since $(a, a, a) \in \kappa$, we may use [S] to deduce that $\sigma_a\sigma_a\sigma_a = \sigma_b \in S$. Then a, b are axes of σ_b, and so by (a) we have $a = b$ or $\sigma_b = 1$. The first case immediately implies $\sigma_a\sigma_a = 1$, as required. In the second case we have $(b, a, a) \in \kappa$ and [S] yields that there exists c such that $\sigma_b\sigma_a\sigma_a = \sigma_c$. Now a is an axis of σ_b, hence a, c are both axes of σ_c. Thus $a = c$ or $\sigma_c = 1$. But both these alternatives imply at once that $\sigma_a^2 = 1$.

(c): Let a, b be distinct lines of L. By (1), $\sigma_a\sigma_b \neq 1$; by (b) $\sigma_a = \sigma_a^{-1}$, hence $\sigma_b \neq \sigma_a$.

(d): Let a, b be arbitrary lines of L. Then $(a, b, a) \in \kappa$, and so by [S] we have $\sigma_a\sigma_b\sigma_a \in S$, and $b\sigma_a$ is an axis of $\sigma_a^{-1}\sigma_b\sigma_a$. By (a) either $\sigma_a\sigma_b\sigma_a = \sigma_{b\sigma_a}$ or $\sigma_a\sigma_b\sigma_a = 1$. In the latter case we have $\sigma_b = 1$, and $\sigma_a\sigma_b\sigma_a = \sigma_b$. If $b \neq b\sigma_a$, then by (c) $\sigma_{b\sigma_a} \neq 1$, hence $\sigma_a\sigma_{b\sigma_a}\sigma_a = \sigma_{b\sigma_a\sigma_a} = \sigma_b$, implying $\sigma_a\sigma_b\sigma_a = \sigma_{b\sigma_a}$, as claimed.*

(e): Using (b), (c), and (d) yields the following equivalent statements: $\sigma_a\sigma_b = \sigma_b\sigma_a \Leftrightarrow \sigma_a\sigma_b\sigma_a = \sigma_b \Leftrightarrow \sigma_{b\sigma_a} = \sigma_b \Leftrightarrow b\sigma_a = b$.

(f): Suppose $\sigma_a\sigma_b\sigma_c$ is an involution. If $(a, b, c) \in \kappa$, then by [S] $\sigma_a\sigma_b\sigma_c \in S$, as claimed. Thus we assume now that $(a, b, c) \notin \kappa$. Then both

---

* Since $\sigma_{b\sigma_a} \neq 1$, and $\sigma_b = 1$, this cannot occur.

a, b, c, and $L(a, b)$, $L(a, c)$ and $L(b, c)$ are pairwise distinct. We first show

(3) $\sigma_a\sigma_b$, $\sigma_a\sigma_c$, $\sigma_b\sigma_c$ are involutions.

Because of symmetry it suffices to show this for $\sigma_a\sigma_b$. Since $a \neq b$, we have $\sigma_a\sigma_b \neq 1$ by (1). Using (d) we also obtain $\sigma_{b\sigma_a}\sigma_{c\sigma_a} = \sigma_a\sigma_b\sigma_a\sigma_a\sigma_c\sigma_a = (\sigma_a\sigma_b\sigma_c)^2\sigma_c\sigma_b = \sigma_c\sigma_b$, hence $\sigma_{b\sigma_a}\sigma_{c\sigma_a}\sigma_b \in S$. By [S] this implies $(b\sigma_a, c\sigma_a, b) \in \kappa$, hence $b \in L(b\sigma_a, c\sigma_a) = L(b, c)\sigma_a$. Since $L(a, b)$ lies on a, we have $L(a, b)\sigma_a = L(a, b)$, hence $b\sigma_a \in L(a, b)$. Thus $b\sigma_a, b \in L(a, b)$, $L(b, c)$, and by assumption $L(a, b) \neq L(b, c)$, which implies that $b\sigma_a = b$. Hence by (e) we have $(\sigma_a\sigma_b)^2 = 1$, and thus $\sigma_a\sigma_b$ is an involution. This proves (3).

Combining (e) with (3), we see that each of the lines a, b, c, hence each of the vertices of the trilateral a, b, c, remains fixed under each reflection σ_a, σ_b, and σ_c. Since $(a, b, c) \notin \kappa$, we have therefore proved

(4) L(a, b) *is linewise fixed under* σ_c; L(a, c) *is linewise fixed under* σ_b, *and* L(b, c) *is linewise fixed under* σ_a.

Now let $C := L(a, b)$. We prove

(5) C *is linewise fixed under* $\sigma_a\sigma_b$.

For let x be a line of C. Then $(a, b, x) \in \kappa$, and so by [S] $\sigma_a\sigma_b\sigma_x \in S$. Using (d), (b), and (3), we then have $\sigma_{x\sigma_a\sigma_b} = \sigma_b\sigma_a\sigma_x\sigma_a\sigma_b = \sigma_b\sigma_a\sigma_b\sigma_a\sigma_x = \sigma_x$, hence by (c) $x\sigma_a\sigma_b = x$. This proves (5).

Similarly, we define $A := L(b, c)$ and $B := L(a, c)$, and we have

(6) A *is linewise fixed under* $\sigma_b\sigma_c$; B *is linewise fixed under* $\sigma_a\sigma_c$.

Next we show

(7) *If* σ_a, $\sigma_b \neq 1$, *then* c *is an axis of* $\sigma_a\sigma_b$.

By (5) all the points on c that are connected to C remain fixed under $\sigma_a\sigma_b$. Suppose c is not an axis of $\sigma_a\sigma_b$. Since C is Δ-connected, this implies that there exist exactly two points D, E on c not connected with C such that $D\sigma_a\sigma_b = E$. Since $C\sigma_a = C$, we also know that $D\sigma_a$ is a point on c which is not connected with C. Hence either $D\sigma_a = D$ or $D\sigma_a = E = D\sigma_a\sigma_b$. By symmetry in a, b we can assume without loss of generality that $D\sigma_a = D$, and so $D\sigma_a\sigma_b = D\sigma_b$. Then D is linewise fixed by σ_a. For $r \in D$, $r \neq c$ let $F := L(r, b)$. Then F is linewise fixed under σ_a. Now $\sigma_a \neq 1$, and so by (a) c is not an axis of σ_a. But since $c \notin F$, we know that every point on c that is connected with F is fixed under σ_a. Hence there exist two points H, H' on c, neither of which is connected with F, and $H\sigma_a = H'$. Clearly, the six points A, B, D, E, H, H' on c are pairwise distinct.

Since F is Δ-connected, it is connected to B by some line s. Let $t \neq c$ be a line of H; then $t \notin F$ [since F, H are not connected, and the points $L(t, y)$ for $y \in F$ are mutually distinct]. Suppose $t \notin L(a, r\sigma_b)$. Then D is

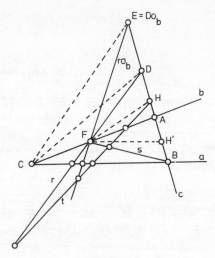

Figure 1

connected to one of the points $L(t, r\sigma_b)$, $L(t, s)$, $L(t, b)$ by some line m (since D is Δ-connected). Since $m\sigma_a = m$, and $y\sigma_a = y$ for all $y \in F$, we know that this point is fixed under σ_a. Also $L(t, a)$ is fixed under σ_a, hence t is fixed under σ_a. But then $H = L(t, c)$ is fixed under σ_a, contradicting $H\sigma_a = H' \neq H$.

So we assume $t \in L(a, r\sigma_b)$, implying that $t \notin L(a, r)$. We then proceed as above, using the point E instead of D, and r instead of $r\sigma_b$. Hence the assumption that c is not an axis of $\sigma_a\sigma_b$ leads to a contradiction, proving (7).

By similar reasoning we have the following

(8)   If $\sigma_b, \sigma_c \neq 1$,   then   a is an axis of $\sigma_b\sigma_c$;
      if $\sigma_a, \sigma_c \neq 1$,   then   b is an axis of $\sigma_a\sigma_c$.

We now proceed as follows: by symmetry in a, b, c we may assume without loss of generality by (c) that $\sigma_a, \sigma_b \neq 1$. We claim that $\sigma_c = 1$. For suppose that $\sigma_c \neq 1$. Then $\sigma_a\sigma_b\sigma_c = (\sigma_a\sigma_b)\sigma_c = (\sigma_a\sigma_c)\sigma_b = (\sigma_b\sigma_c)\sigma_a$. By (7) and (8) we would conclude that the collineation $\sigma_a\sigma_b\sigma_c$ has axes a, b, c, implying that $\sigma_a\sigma_b\sigma_c = 1$, in contradiction to the assumption that $\sigma_a\sigma_b\sigma_c$ is an involution. Thus $\sigma_c = 1$. Further, we claim that $|A| = 2$. To show it suppose A contained some $b' \neq b, c$; then $(b, b', c) \in \kappa$, and so by [S] $\sigma_b\sigma_{b'}\sigma_c = \sigma_b\sigma_{b'} \in S$, which implies by (b) that $(\sigma_b\sigma_{b'})^2 = 1$. Since $b'\sigma_a = b'$, we then use (e) to conclude that $(\sigma_a\sigma_{b'})^2 = 1$, hence $\sigma_a\sigma_b\sigma_{b'} = \sigma_{b'}\sigma_b\sigma_a$. Thus $\sigma_a\sigma_b\sigma_{b'}$ is involutory, or $\sigma_a\sigma_b\sigma_{b'} = 1$. But the former yields $\sigma_a, \sigma_b, \sigma_{b'} \neq 1$ (since $b' \neq c$ and $\sigma_c = 1$), and by an argument similar to the above one we also obtain $\sigma_a\sigma_b\sigma_{b'} = 1$. But then $\sigma_a\sigma_b = \sigma_{b'}$, and so by (7) we see that $\sigma_{b'}$ has distinct axes b', c, and we infer by (a) that $\sigma_{b'} = 1$, a contradiction. Hence $|A| = 2$, and similarly, $|B| = 2$.*

As in the proof of (a), this implies that $|x| \leq 4$ for all lines $x \neq c$, and $|X| \leq 4$ for all points X. By assumption, we know there exists a line d such that no three of a, b, c, d are concurrent. Let $D := L(c, d)$, $E := L(c, d\sigma_a)$, and $e := d\sigma_a\sigma_b$. We observe that $d\sigma_a \neq d$, for otherwise d is contained in the three points D, $L(a, d)$, $L(b, d)$, each of which would be fixed by σ_a; as $|d| \leq 4$, this would imply that d is an axis of σ_a, which, together with $d \neq a$ and $\sigma_a \neq 1$, gives a contradiction. Similarly, $d\sigma_b \neq d$. Now

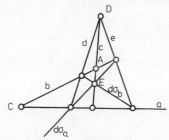

Figure 2

$L(c, d\sigma_b) = L(c, d\sigma_a)\sigma_a\sigma_b = L(c, d\sigma_a)$ by (7), hence $d\sigma_b \in L(c, d\sigma_a)$. Similarly, we obtain $e \in D$, and thus the lines a, b, d, $d\sigma_a$, $d\sigma_b$, e form a complete quadrilateral with diagonal points C, D, and E. Since $d \neq c$, we have by (c) that $\sigma_d \neq 1$. As in the above, we conclude from $\sigma_d\sigma_e\sigma_c \in S$ that $(\sigma_d\sigma_e)^2 = 1$, and so by (e) we have $e\sigma_d = e$. Also, we have $\sigma_d\sigma_c = \sigma_c\sigma_d$, and thus $c\sigma_d = c$, again by (e). Since $|D| \leq 4$, we therefore know that D is linewise fixed by σ_d. This implies that $x\sigma_d \neq x$ for all lines $x \notin D$, as otherwise x would perforce be an axis of σ_d, and this, together with (a) and $\sigma_d \neq 1$, would imply $x = d \in D$. Hence D is the only point of e that remains fixed by σ_d, hence $|e| - 1$ is even, which implies that $|e| = 3$. This entails $b\sigma_d = d\sigma_b$, hence $A\sigma_d = E$. But c, $d\sigma_a$, $d\sigma_b$ are pairwise distinct, hence $|E| \geq 3$. Thus we have arrived at the contradiction $2 = |A| = |A\sigma_d| = |E| \geq 3$.

The assumption $(a, b, c) \notin \kappa$ is therefore wrong, hence we have proved (f). This completes the proof of the lemma. ■

We remark that (e) in Lemma 3.2 enables us to introduce the concept of orthogonality for the complete metric planes: Two lines a, b of a complete metric plane are said to be *orthogonal* if $\sigma_a\sigma_b$ is involutory.

Obviously, the relation

$$\omega := \{(a, b) \mid \sigma_a\sigma_b \text{ is involutory}\}$$

---

* Observe that this concludes the proof of (f) if we assume that every point lies on at least three lines.

is irreflexive and symmetric. To indicate the dependence of ω on the map Φ we write ω_Φ instead of ω. Thus, if (L, κ, Φ) is a complete metric plane, then (L, κ, ω_Φ) is an incidence structure with orthogonality.

A thorough characterization of the complete metric planes is given by Main Theorem 6.31 on complete metric planes in Chapter 6, Section 9.

6. METRIC PLANE OVER A METRIC VECTOR SPACE

Throughout this section let (V, Q) be a three-dimensional metric vector space over a field K and $Q \neq 0$. Mainly, we must exclude two exceptional cases which do not satisfy the following condition:

$$\text{If} \quad K = GF(2), \quad \text{then} \quad \text{Ind}\,(V, Q) = 0 \qquad \qquad [E]$$

There exist exactly two metric vector spaces (V, Q) over $GF(2)$ such that $\text{Ind}\,(V, Q) \neq 0$ (and therefore $\text{Ind}\,(V, Q) = 1$).

In the notations of Chapters 1 and 2 we consider the incidence structure $I(V, Q)$ over the three-dimensional metric vector space (V, Q) over a field K. We recall that the set $L(V, Q)$ of lines of this incidence structure consists of all the nonisotropic one-dimensional subspaces of V, hence of all the nonisotropic lines in the projective coordinate incidence structure $I(V)$. Moreover, the concurrency relation $\kappa(V, Q)$ is the restriction of the concurrency relation in the projective coordinate incidence structure $I(V)$ to the subset $L(V, Q)$ of lines.

Obviously, any collineation α of $I(V)$ mapping $L(V, Q)$ onto itself induces a collineation $\bar{\alpha}$ of $I(V, Q)$, and we only have to consider the restriction of α to $L(V, Q)$. This restriction determines the collineation α uniquely, since we can prove:

LEMMA 3.3. *Let* (V, Q) *be a three-dimensional metric vector space satisfying* $[E]$ *and* $Q \neq 0$. *Then for any two collineations* α, β *of* $I(V)$

$$\alpha\,|_{L(V,Q)} = \beta\,|_{L(V,Q)} \quad \textit{implies} \quad \alpha = \beta$$

PROOF.   By assumption, α and β coincide for all nonisotropic lines in $I(V)$. Comparing Proposition 2.4 and observing $[E]$, we see that any point in $I(V)$ is incident with at least two nonisotropic lines, the only exceptions being the points A and B, C in the cases (c) and (e) of Proposition 2.4, respectively, which make up the set $L(V)\backslash L(V, Q)$. Hence α and β coincide for all points $\neq A$, B, C in $I(V)$. However, this implies that α and β coincide for all lines in $I(V)$, and therefore we have $\alpha = \beta$. ∎

By the Appendix, any linear map of V onto V induces a collineation of the projective incidence structure $I(V)$. Especially, any linear map σ_A in $S(V, Q)$, and therefore any element α in $O(V, Q)$, induces a collineation $\bar{\alpha}$ of $I(V)$. By (2) in Chapter 1, Section 4, α maps nonisotropic vectors onto nonisotropic vectors, and thus the restriction of $\bar{\alpha}$ to $L(V, Q)$ is a collineation of $I(V, Q)$. By Lemma 3.3 we are justified to denote the restriction of $\bar{\alpha}$ to $L(V, Q)$ by the same symbol $\bar{\alpha}$.

LEMMA 3.4. *Let* (V, Q) *be a three-dimensional metric vector space satisfying* $[E]$ *and* $Q \neq 0$. *For* $\alpha \in O(V, Q)$ *we denote by* $\bar{\alpha}$ *the restriction of the collineation of* $I(V)$ *induced by* α *to* $L(V, Q)$. *Then for* $\alpha, \beta \in O(V, Q)$:

$$\bar{\alpha} = \bar{\beta} \quad \text{implies} \quad \alpha = \beta$$

PROOF. Let $\alpha, \beta \in O(V, Q)$ and $\bar{\alpha} = \bar{\beta}$. We denote by α', β' the collineations of $I(V)$ induced by α, β. Then Lemma 3.3 yields that $\alpha' = \beta'$. By the Appendix there exists an element s in the field K such that $\alpha = s\beta$, and by Lemma 1.10 we conclude $s = 1$ and therefore $\alpha = \beta$. ∎

Now let $\sigma_A \in S(V, Q)$. Then the restriction $\bar{\sigma}_A$ to $L(V, Q)$ of the collineation of $I(V)$ induced by σ_A is a collineation of $I(V, Q)$ with the axis $\langle A \rangle$, since the definition of σ_A yields that for any line $\langle X \rangle$ in V the lines $\langle A \rangle$, $\langle X \rangle$, $\langle X \rangle \bar{\sigma}_A$ are concurrent. Let $S'(V, Q)$ denote all the collineations of $I(V, Q)$ induced by any element in $S(V, Q)$. Then the map

$$\mu : \begin{cases} S(V, Q) \to S'(V, Q) \\ \sigma_A \mapsto \bar{\sigma}_A \end{cases}$$

is a bijection provided that (V, Q) satisfies $[E]$ and $Q \neq 0$: Obviously, it is surjective, and the injectivity follows immediately from Lemma 3.4.

The group generated by $S'(V, Q)$ may be denoted by $O'(V, Q)$. The map μ induces an isomorphism $\bar{\mu}$ of $(O(V, Q), S(V, Q))$ onto $(O'(V, Q), S'(V, Q))$ if (V, Q) fulfills the condition $[E]$ and $Q \neq 0$: We put

$$\bar{\mu}(\sigma_A \sigma_B \cdots \sigma_X) := \bar{\sigma}_A \bar{\sigma}_B \cdots \bar{\sigma}_X$$

We consider the map*

$$\Phi(V, Q) : \begin{cases} L(V, Q) \to S'(V, Q) \\ \langle A \rangle \mapsto \bar{\sigma}_A \end{cases}$$

It is a mapping of $L(V, Q)$ into the set of the axial collineations of $I(V, Q)$. By our remark above and by Lemma 1.11, $\Phi(V, Q)$ is a bijective map if dim $V^\perp \leq 2$.

---

*Since μ is a bijection of $S(V, Q)$ onto $S'(V, Q)$, we are allowed to denote the mapping defined in Chapter 1, Section 4, and the following map by the same symbol, although the ranges of the two mappings are distinct.

Let (V, Q) be any three-dimensional metric vector space satisfying [E] and dim $V^\perp \le 2$. Then we call the triplet

$$M(V, Q) := (L(V, Q), \kappa(V, Q), \Phi(V, Q))$$

the *metric plane over the metric vector space* (V, Q) and prove:

PROPOSITION 3.5. *Let* (V, Q) *be a three-dimensional metric vector space satisfying* [E] *and dim* $V^\perp \le 1$. *Then the metric plane* $M(V, Q)$ *over* (V, Q) *is a complete metric plane.*

PROOF. By Lemma 2.3 and Lemma 2.5 $I(V, Q)$ is a Δ-connected incidence structure, which contains a quadrilateral.

Thus we only have to prove that $M(V, Q)$ fulfills [S]. To prepare this we compute the product of three linear mappings σ_A, σ_B, σ_C in $S(V, Q)$ (compare Chapter 1, Section 4):

$$X\sigma_A\sigma_B\sigma_C = (-X + (X \circ A)A)\sigma_B\sigma_C$$
$$= (X - (X \circ B)B)\sigma_C + (X \circ A)(-A + (A \circ B)B)\sigma_C$$
$$= -X + (X \circ C)C - (X \circ B)(-B + (B \circ C)C) +$$
$$(X \circ A)(A - (A \circ C)C)$$
$$+ (X \circ A)(A \circ B)(-B + (B \circ C)C)$$

Simplifying this, we obtain

$$X + X\sigma_A\sigma_B\sigma_C = (X \circ A)A + (X \circ B - (X \circ A)(A \circ B))B$$
$$+ (X \circ C - (X \circ B)(B \circ C) - (X \circ A)(A \circ C) + (X \circ A)(A \circ B)(B \circ C))C \quad (*)$$

Now we prove [S].

(A) Let $\bar\sigma_A\bar\sigma_B\bar\sigma_C$ be a reflection $\bar\sigma_D$ in $S'(V, Q)$. Then by Lemma 3.4 we know $\sigma_A\sigma_B\sigma_C = \sigma_D$. By Lemma 1.1 we have dim $\langle D\rangle^\perp \ge 2$. Thus we can choose a vector X such that $X \in \langle D\rangle^\perp \setminus V^\perp$.

By definition $X + X\sigma_A\sigma_B\sigma_C = X + X\sigma_D = O$; thus by $(*)$ A, B, C are linearly dependent or the coefficients of A, B, C in $(*)$ are all 0. In the latter case we obtain successively $X \circ A = 0$ and $X \circ B = 0$ and $X \circ C = 0$, hence $g(X, A) = g(X, B) = g(X, C) = 0$, and therefore A, B, $C \in \langle X\rangle^\perp$. Since $X \notin V^\perp$, by Lemma 1.1 dim $\langle X\rangle^\perp = 2$, and A, B, C are linearly dependent in either case. Hence the lines $\langle A\rangle$, $\langle B\rangle$, $\langle C\rangle$ are concurrent.

(B) Now let $\langle A\rangle$, $\langle B\rangle$, $\langle C\rangle$ be three concurrent lines in $L(V, Q)$. If $\langle A\rangle = \langle B\rangle$, then $\bar\sigma_A\bar\sigma_B\bar\sigma_C = \bar\sigma_C \in S'(V, Q)$. If $\langle B\rangle = \langle C\rangle$, then $\bar\sigma_A\bar\sigma_B\bar\sigma_C = \bar\sigma_A \in S'(V, Q)$. If $\langle A\rangle = \langle C\rangle$, then an easy computation shows from $(*)$ that $\sigma_A\sigma_B\sigma_C = \sigma_{B\sigma_A}$, hence $\bar\sigma_A\bar\sigma_B\bar\sigma_C = \bar\sigma_{B\sigma_A} \in S'(V, Q)$.

Thus we may assume $\langle A\rangle$, $\langle B\rangle$, $\langle C\rangle$ are mutually distinct lines. Then A, B are linearly independent, and we may assume without loss of generality that $C = A + B$. Using (*), we obtain

$$X + X\sigma_A\sigma_B\sigma_C = (X \circ A + X \circ C - (X \circ B)(B \circ C) - (X \circ A)(A \circ C)$$
$$+ (X \circ A)(A \circ B)(B \circ C))A + (X \circ B - (X \circ A)(A \circ B)$$
$$+ X \circ C - (X \circ B)(B \circ C) - (X \circ A)(A \circ C)$$
$$+ (X \circ A)(A \circ B)(B \circ C))B$$

Multiplying this equation by $Q(A)Q(B)Q(C)$, compute that

$$Q(A)Q(B)Q(C)(X + X\sigma_A\sigma_B\sigma_C)$$
$$= g(X, A)[Q(B)(Q(C) + Q(A) - g(A, C)) + g(A, B)g(B, C)]A$$
$$+ g(X, B)[Q(A)Q(B) - Q(A)g(B, C)]A$$
$$+ g(X, B)[Q(A)(Q(C) + Q(B) - g(B, C))]B$$
$$+ g(X, A)[Q(A)Q(B) - Q(C)g(A, B) - Q(B)g(A, C)$$
$$+ g(A, B)g(B, C)]B$$

Since $g(B, C) = g(B, B) + g(B, A) = 2Q(B) + g(A, B)$ [compare equation 1 in Chapter 1, Section 1] and analogously $g(A, C) = 2Q(A) + g(A, B)$, we can apply equations 1 and 2 in Chapter 1, Section 1, to obtain

$$Q(A)Q(B)Q(C)(X + X\sigma_A\sigma_B\sigma_C) = g(X, A)(Q(B) + g(A, B))^2 A$$
$$- g(X, B)Q(A)(Q(B) + g(A, B))A$$
$$- g(X, A)Q(A)(Q(B) + g(A, B))B$$
$$+ g(X, B)Q(A)^2 B$$
$$= g(X, (Q(B) + g(A, B))A - Q(A)B)$$
$$\times [(Q(B) + g(A, B))A - Q(A)B]$$
$$= g(X, D)D$$

if we put $D := (Q(B) + g(A, B))A - Q(A)B$. Furthermore, we compute that $Q(D) = Q(A)Q(B)Q(C)$; then we have $Q(D) \neq 0$ and

$$X\sigma_A\sigma_B\sigma_C = -X + \frac{g(X, D)}{Q(D)} D = X\sigma_D \quad \text{for all} \quad X \text{ in } V$$

Hence $\bar{\sigma}_A\bar{\sigma}_B\bar{\sigma}_C$ is the reflection $\bar{\sigma}_D$ in $S'(V, Q)$, and [S] is proved.
This completes the proof of Proposition 3.5. ∎

7. DEFINITION OF EUCLIDEAN AND NON-EUCLIDEAN COORDINATE PLANES

Let (V, Q) by any three-dimensional metric vector space such that $Q \neq 0$. According to the fact that there are only five types separated by the possible numbers of dim rad (V, Q) and Ind (V, Q) (compare Proposition 2.4), we have five types of incidence structures over a metric vector space. By our elaborations in Sections 1 and 3 we are justified to introduce the following names for each of the five types:

The incidence structure $I(V, Q) = (L(V, Q), \kappa(V, Q))$ over a three-dimensional metric vector space (V, Q) such that dim $V^{\perp} \leq 2$ is called

	Ind $V, Q)$	dim rad (V, Q)
an *elliptic coordinate plane*, if	0	0
a *Euclidean coordinate plane*, if	0	1
a *Strubecker coordinate plane*, if	0	2
a *hyperbolic-metric coordinate plane*, if	1	0
a *Minkowskian coordinate plane*, if	1	1

For reasons that become apparent later, we also call the triplet $(L(V, Q), \kappa(V, Q), \omega(V, Q))$ and also the triplet

$$(L(V, Q), \kappa(V, Q), \Phi(V, Q))$$

by the same name as $I(V, Q)$. For the definition of $\omega(V, Q)$ and $\Phi(V, Q)$ compare Sections 3 and 6.

An axiomatic-synthetic definition of an elliptic plane, a Euclidean plane, a Strubecker plane, a hyperbolic-metric plane, and a Minkowskian plane is given in Chapter 6.

4

S-GROUPS AND S-GROUP PLANES

1. DEFINITION

Let G be a group that is generated by a subset S consisting of involutions.

The elements of S are denoted by lowercase letters. Let J be the set of all involutions in G. Then the pair (G, S) is called an *S-group* if the following axiom is satisfied:

AXIOM S. $a \neq b$ *and* $abx, aby, abz \in J$ *implies that* $xyz \in S$.

Two S-groups (G, S) and (G', S') are said to be *isomorphic* if there is an isomorphism α from G onto G' such that $\alpha(S) = S'$. Moreover, two pairs (G, S) and (G', S') consisting of a group and a subset of the group generating the group are said to be isomorphic if there is an isomorphism α of G onto G' such that $\alpha(S) = S'$.

Before giving the geometric interpretation of this, we wish to introduce the following notation:

$$\alpha^{\beta} := \beta^{-1}\alpha\beta \quad \text{for all} \quad \alpha, \beta \in G.$$

$$M^{N} := \{\alpha^{\beta} \mid \alpha \in M, \beta \in N, \quad \text{and} \quad M, N \quad \text{subsets of} \quad G\}$$

If $M = \{\alpha\}$, we write α^{N} instead of $\{\alpha\}^{N}$. Analogously, we replace M^{N} by M^{β} if $N = \{\beta\}$. If n is a positive integer and M is a subset of G, we write

$$M^{n} := \{\alpha_1 \cdots \alpha_n \mid \alpha_1, \ldots, \alpha_n \in M\}$$

2. THE GROUP PLANE OF AN S-GROUP

Let (G, S) be an S-group. The ternary relation κ defined by

$$(a, b, c) \in \kappa \Leftrightarrow abc \in J \qquad (*)$$

is reflexive and symmetric in any group. It is transitive by Axiom S: If $a \neq b$ and $abc, abd \in J$, then $aba, abc, abd \in J$, hence $acd \in S \subset J$. Thus κ is a ternary equivalence relation, and (S, κ) is an incidence structure provided S contains at least two elements. To emphasize the dependence of κ on the S-group, we sometimes write $\kappa(G, S)$ instead of κ.

Given any S-group (G, S) we call the incidence structure $(S, \kappa(G, S))$ the *group plane* of the S-group and denote it by $E(G, S)$. We recall that the elements in S are the lines, and that three lines a, b, c in the group plane $E(G, S)$ are concurrent if $abc \in J$. Moreover, a point is a set of lines*

$$S(ab) := \{x \mid abx \in J\} \quad \text{for} \quad a \neq b$$

Points are denoted by capital letters. If x, y, z are elements in a point, then the lines x, y, z are concurrent (compare Chapter 2, Section 2).

If we interchange the roles of the lines and the points of the group plane of an S-group (G, S), we arrive at the concept of the dual group plane of an S-group: Let P denote the set of all the points of $E(G, S)$; then the triplet $(S, P, \in)$ is called the *dual group plane* of (G, S), denoted by $D(G, S)$. The elements of S are called the *points* of $D(G, S)$, and the elements of P are called the *lines* of $D(G, S)$. A point x of $D(G, S)$ is said to be *incident* with the line Y of $D(G, S)$, if $x \in Y$.

There is a natural concept of orthogonality in the group plane of an S-group (G, S): namely the binary relation

$$(a, b) \in \omega \Leftrightarrow ab \in J \qquad (**)$$

This relation is irreflexive and symmetric. To indicate the dependence of ω on the S-group (G, S), we sometimes write $\omega(G, S)$ instead of ω. The triplet $(S, \kappa(G, S), \omega(G, S))$ is an incidence structure with orthogonality, which sometimes is also called the group plane of the S-group. If in the following we speak in terms of orthogonality for an S-group plane (i.e., a group plane of an S-group), we always mean the orthogonality in the sense of the orthogonality relation $\omega(G, S)$.

Given any two points A, B in any S-group plane such that $a \perp b$ for all a in A and all b in B, we call the point A *orthogonal* to the point B, and we write $A \perp B$. Obviously, we have $A \cap B = \emptyset$. If not $A \perp B$, we write $A \not\perp B$.

A triplet a, b, c of lines in an S-group plane $E(G, S)$ is called a *polar triangle*,[†] if $abc = 1$. Obviously, any two distinct lines in a polar triangle are orthogonal to each other. A line z of an S-group plane $E(G, S)$ is called a *central line* if z is contained in the center of the group G. Clearly,

---

* We have dropped the comma in $S(a, b)$ for brevity. By our notation $S(\alpha)$ is defined if $\alpha \neq 1$ and $\alpha \in S^2$.
† Strictly speaking, we should call the triplet a, b, c a *polar trilateral*.

any line distinct from z in the group plane is orthogonal to the central line z.

3. ELEMENTARY PROPERTIES

We wish to prove some elementary properties of S-groups. At times we use the geometric interpretation. Without assuming Axiom S, we can prove

① *If* ab = ba *and* ac = ca, *then* abc = cba *is equivalent to* bc = cb.

PROOF. $abc = cba \Leftrightarrow bac = cba \Leftrightarrow bca = cba \Leftrightarrow bc = cb.$ ∎

For the following we use Axiom S.

② abc ∈ J ⇔ abc ∈ S.

PROOF. If $abc \in J$, then either $a = b$ and $abc = c \in S$, or $a \neq b$ and aba, abb, $abc \in J$ imply $abc \in S$ by Axiom S. If $abc \in S$, then of course $abc \in J$, since $S \subseteq J$. ∎

By ② we may define the concurrency relation in the group-plane $E(G, S)$ of the S-group as follows:

$$(a, b, c) \in \kappa \Leftrightarrow abc \in S \qquad (*)$$

The points can be defined by

$$S(ab) := \{x \mid abx \in S\} \quad \text{for} \quad ab \neq 1$$

③ $S^{\alpha} = S$ *for all* $\alpha \in G$.

PROOF. Since α is a product of elements of S, it suffices to consider the case that α is an element a of S. In this case $x^{a} = axa \in J$, and by ② $x^{a} \in S$ for all x in S. Thus $S^{\alpha} \subset S$ for all $\alpha \in G$ and therefore $S^{\alpha^{-1}} \subset S$. However, this implies $S \subset S^{\alpha}$, hence $S^{\alpha} = S$. ∎

④ $a \neq b$ *and* abc, abd ∈ S *implies that* acd ∈ S.

PROOF. This follows immediately from ② and the transitivity of κ. ∎

⑤ a ∈ A, A^b *implies* b ∈ A *or* b ∉ A *and* a ∈ b⊥.

PROOF. If $b \notin A$, then $b \neq a$ and $a, a^{b} \in A$, $S(ab)$. Since $b \notin A$, we know $A \neq S(ab)$. Hence $a = a^{b}$ and $a \neq b$. Thus $a \in b^{\perp}$. ∎

Note that A^b is a point, since $A^b = S(xy)^b = S(x^b y^b)$.

⑥ $A^b = A$ *implies* $b \in A$ *or* $A \subset b^\perp$.

PROOF. If $b \notin A$, then $x \in A$ implies that $x \in A^b$, and by ⑤ $x \in b^\perp$. Hence $A \subset b^\perp$. ∎

⑦ $x, y \in a^\perp$ *and* $xay \notin J$ *implies* $S(xy) \subset a^\perp$.

PROOF. If $xay \notin J$, then $x \neq y$ and $S(xy)^a = S(x^a y^a) = S(xy)$. Since $a \notin S(xy)$, we conclude from ⑥ that $S(xy) \subset a^\perp$. ∎

⑧ $x, y, z \in A$ *implies* $xyz \in A$.

PROOF. If $x = y$, then $xyz = z \in A$. If $x \neq y$, then $A = S(yx)$ and $(yx)xyz = z \in J$, thus $xyz \in S(yx) = A$. ∎

⑨ A^2 *is an abelian group for any point* A.

PROOF. Let ab, cd be any elements of A^2. Then $(ab)(cd)^{-1} \in A^2$, since $abd \in A$. Thus A^2 is a group. From $(ab)(cd) = (abc)d = cbad = (cd)(ab)$ we deduce that A^2 is abelian. ∎

Statements ① to ⑨ are used very often without explicit reference.

4. REFLECTIONS AND MOTIONS

Let α be any element of G; then by ③ $S^\alpha = S$. Thus we may consider the map

$$\alpha^* : \begin{cases} S \to S \\ x \mapsto x^\alpha \end{cases}$$

This map is an orthogonal collineation of $E(G, S) = (S, \kappa, \omega)$, since $(a, b, c) \in \kappa \Leftrightarrow abc \in J \Leftrightarrow (abc)^\alpha \in J \Leftrightarrow a^\alpha b^\alpha c^\alpha \in J \Leftrightarrow (a^\alpha, b^\alpha, c^\alpha) \in \kappa$ and $(a, b) \in \omega \Leftrightarrow ab \in J \Leftrightarrow (ab)^\alpha \in J \Leftrightarrow a^\alpha b^\alpha \in J \Leftrightarrow (a^\alpha, b^\alpha) \in \omega$.

We call α^* a *motion* of the group plane $E(G, S)$. If α is an element g in S, then g^* is a reflection in the line g: $g^{*2} = 1$; $g^g = g$; $aga^g = g \in J$; thus $(a, g, a^g) \in \kappa$ for all a in S; $a^g = a$ and $a \neq g \Leftrightarrow ag \in J \Leftrightarrow (a, g) \in \omega$.

Axiom S reads: If x, y, z are concurrent lines, the product $x^* y^* z^*$ of reflections in the lines x, y, z is a reflection u^* in a line.

We observe that if z is a central line, then the reflection z^* in the line z is the identity.

We recall that if $A = S(ab)$ is a point in $E(G, S)$ and α^* is a motion of $E(G, S)$, then

$$A\alpha^* = S(ab)\alpha^* = S(ab)^\alpha = S(a^\alpha b^\alpha)$$

Thus we have for any line a and any point B

$$a \in B \Leftrightarrow a^\alpha \in B^\alpha \Leftrightarrow a\alpha^* \in B\alpha^*$$

We put $G^* := \{\alpha^* \mid \alpha \in G\}$ and $S^* := \{a^* \mid a \in S\}$. The group G^* is called the *group of motions* of the S-group plane $E(G, S)$. Clearly, if we denote by $Z(G)$ the center of the group G, we have $G^* \cong G/Z(G)$.

Remark. Throughout this book the symbol α^* for an element α of an S-group always denotes the motion defined above.

5. SOME METRIC CONFIGURATION THEOREMS FOR S-GROUP PLANES

LEMMA 4.1 (Lemma of the nine lines).* *Let (G, S) be any S-group and let α_i, β_k $(i, k = 1, 2, 3)$ be elements of G such that $\alpha_1 \neq \alpha_2$ and $\beta_1 \neq \beta_2$. If $\alpha_i \beta_k \in S$ for all pairs $(i, k) \neq (3, 3)$, then $\alpha_3 \beta_3 \in S$.*

	β_1	β_2	β_3
α_1	$\times$	$\times$	$\times$
α_2	$\times$	$\times$	$\times$
α_3	$\times$	$\times$	$\square$

PROOF. If $\beta_1 = \beta_3$, the statement is trivial. Let $\beta_1 \neq \beta_3$. Then $\alpha_1\beta_1$, $\alpha_2\beta_1 \in S((\alpha_1\beta_1)^{-1}(\alpha_1\beta_2))$, $S((\alpha_1\beta_1)^{-1}(\alpha_1\beta_3))$ and $\alpha_1\beta_1 \neq \alpha_2\beta_1$. Hence $S((\alpha_1\beta_1)^{-1}(\alpha_1\beta_2)) = S((\alpha_1\beta_1)^{-1}(\alpha_1\beta_3))$. If we denote this point by A, we have $\alpha_3\beta_1$, $\alpha_1\beta_1$, $\alpha_1\beta_3 \in A$ and thus $\alpha_3\beta_3 = \alpha_3\beta_1(\alpha_1\beta_1)^{-1}(\alpha_1\beta_3) \in S$. This proves the lemma. ∎

Let A be any point of an S-group plane. For $a, b \in A$, the mapping

$$\gamma : \begin{cases} A \to A \\ x \mapsto axb \end{cases}$$

is denoted by γ_{ab} (Gegenpaarung). The definition yields the following:

$$\text{If} \quad \gamma := \gamma_{ab}, \quad \text{then} \quad xy = (y\gamma)(x\gamma) \quad \text{for all} \quad x, y \in A \qquad (1)$$

PROOF. $xy = xybb = byxb = bya \cdot axb = ayb \cdot axb = (y\gamma)(x\gamma)$. ∎

---

* Compare Bachmann [2], Section 4.7.

THEOREM 4.2 (Gegenpaarung). *Let* a_1, a_2, a_3 *be three lines in the domain of a mapping* $\gamma := \gamma_{ab}$ *and let* $b_i := a_i\gamma$, $i = 1, 2, 3$. *If* c_1, c_2, c_3 *and* $c_ia_kc_l$ *are lines for any permutation* i, k, l *of* 1, 2, 3 *and* $c_1b_1 \neq c_2b_2$, *then* c_1b_1g, $c_2b_2g \in J$ *implies* $c_3b_3g \in J$ *for some line g.*

PROOF.* Put $a := (a_1a_2a_3)^{c_2}$; $b := c_1a_3c_2$; $c := c_2a_1c_3$; $d := c_2b_2g$. Then by equation 1 $ab = c_2a_1a_2c_1 = c_2b_2b_1c_1 \neq 1$. Hence $a \neq b$ and $abc = (a_2c_1c_3)^{a_1c_2} \in S$ and $abd = (b_1c_1g)^{b_2c_2} \in S$. Thus by ④ $acd = (gb_3c_3)^{cb_2c_2} \in S$, hence $c_3b_3g \in J$. ∎

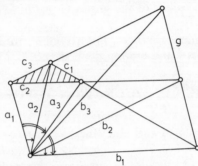

THEOREM 4.3 (Altitudes of a triangle). *Let* $(abc)^2 \neq 1$ *and*

$$au, bv, cw \in J; \quad bcu, cav, abw \in J \tag{2}$$

Then $uvw \in J$.

PROOF.† Put

$$a' := abw; \quad b' := (cw)(abc); \quad c' := (abc)(ua); \quad d' := avc$$

These elements are lines by equation 2, and we have $a'b' = (abc)^2 \neq 1$. Then by ④ $a' \neq b'$ and $a'b'c' = c'^{cba} \in S$ and $a'b'd' = d'^{cba} \in S$ imply $a'c'd' = (uvw)^{cw} \in S$. Hence $uvw \in J$. ∎

If the assumption $(abc)^2 \neq 1$ in the theorem of the altitudes of a triangle fails to hold, then either $abc = 1$ and a, b, c is a polar triangle (a set of three non-concurrent lines that are pairwise orthogonal), or $abc \in J$ and a, b, c are concurrent lines. In the first case the statement of the theorem is not true, since a, b, c are altitudes of the triangle a, b, c and are not concurrent. In the second case either two of the lines, say a, b, are distinct, or all the lines coincide. If $a \neq b$, then the statement of the theorem is trivially correct, since $c \in S(ab)$ and thus $u, v, w \in S(ab)$. Finally, if $a = b = c$, then

---

* Compare Bachmann [2], Section 4.8, Theorem 12, pp. 69–70.
† The first proof of the theorem of the altitudes of a triangle in the theory of S-groups is due to U. Ott.

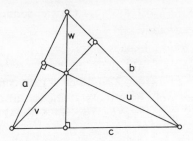

the statement of the theorem may be true or it may be false. If $a^\perp$ is contained in a point, then the assumption holds (compare Theorem 5.4). On the other hand there are S-groups, such that $a^\perp$ is not always contained in a point (for example, S-groups in which a is an element of the center).

6. THE CENTER OF AN S-GROUP

Let (G, S) be any S-group. Then we call the center $Z(G)$ of the group G the *center* of the S-group (G, S).

THEOREM 4.4 (Center of an S-group). *Let* (G, S) *be an S-group and let* Z *be its center. If* S *contains a quadrilateral, then*

(a) $Z \cap S$ *contains no more than one element.*
(b) $Z \cap S^2 = \{1\}$.
(c) $Z \cap S^3 \subset S \cup \{1\}$.
(d) $Z \cap S^4 \cap J = \varnothing$.

PROOF. First we prove

(1) $\alpha, \beta \in J$ *and* $\alpha\beta \in Z$ *implies* $(\alpha\beta)^2 = 1$.

We have $(\alpha\beta)^2 = (\alpha\beta)\alpha\beta = \alpha(\alpha\beta)\beta = 1$. Next we prove (b):

(2) $\alpha \in Z \cap S^2$ *implies* $\alpha = 1$.

If $\alpha \neq 1$, then by (1) $\alpha \in J$. Also $\alpha x = x\alpha = x\alpha^{-1}$ for all x. Hence $x \in S(\alpha)$ or $x = \alpha$ for all x, contradicting the assumption that S contains a quadrilateral.

(3) $Z \cap S$ *contains at most one element.*

If $z, z' \in Z \cap S$, then $zz' \in Z \cap S^2$, hence by (2) $zz' = 1$, implying $z = z'$.

(4) $\alpha \in Z \cap S^3$ *implies* $\alpha \in S$ *or* $\alpha = 1$.

Let α be any element of $Z \cap S^3$, and $\alpha \notin S$. Then $\alpha = abc$ and $\alpha \notin J$. Hence $b \neq c$. Now $bc = a(abc) = (abc)a$ implies $S(bc)^a = S(bc)$, and since $a \notin S(bc)$, we can use ⑥ to conclude that $S(bc) \subset a^\perp$. From this we obtain

$ab \in J$ and thus $\alpha = (ab)c \in J^2 \cap Z$. By (1) we deduce that $\alpha^2 = 1$. Since $\alpha \notin J$, we conclude that $\alpha = 1$.

(5) $\alpha \in Z \cap S^4$ and $\alpha^2 = 1$ implies $\alpha = 1$.

Let $\alpha = abcd \in Z$ and $\alpha^2 = 1$, then $a\alpha = \alpha a = \alpha^{-1}a$, hence $a\alpha = 1$ or $a\alpha \in J$. The first implies $a = \alpha \in Z$ and $bcd = 1$, in contradiction to Corollary 4.6, whose proof only uses the statements (a) and (b). Thus $bcd = a\alpha \in J$, hence $bcd \in S$ and $\alpha = a(bcd) \in S^2 \cap Z$. By (2) we obtain $\alpha = 1$. ∎

Remark.   There exist S-groups satisfying the conditions of Theorem 4.4, whose center contains more than one element $\neq 1$. (Compare Chapter 6.)

COROLLARY 4.5.   *In an S-group plane there is at most one central line, provided there exists a quadrilateral.*

COROLLARY 4.6.   *If an S-group plane contains a quadrilateral, then at most one of the following statements is valid:*

P.   *There exists a polar triangle.*
Z.   *There exists a central line.*

PROOF.   Assuming the S-group plane contains a polar triangle a, b, c and a central line z, then a, b, c are not concurrent, and so we may assume without loss of generality that $z \notin S(ab)$. By $abz = zab = zba$, and $abz \notin J$, we deduce $abz = 1$. Thus $z \in S^2$, contradicting Theorem 4.4(b). ∎

COROLLARY 4.7.   *Let (G, S) be any S-group such that S contains a quadrilateral. Then $S \cup S^2 \cup S^3$ contains at most one element $\neq 1$ belonging to the center of G. If there is such an element, it is an element of S.*

COROLLARY 4.8.   *If $E(G, S)$ is an S-group plane containing a quadrilateral, then the map $\Phi : a \mapsto a^*$ of S onto S^* is injective.*

PROOF.   Let a, b be two lines in $E(G, S)$ such that $a^* = b^*$. Then $x^a = x^b$ and therefore $x^{ab} = x$ for all x in S. This yields $ab \in Z(G)$, hence $ab = 1$ by Theorem 4.4(b). Thus $a = b$, and Φ is injective. ∎

7. S-GROUP PLANES AND S-PLANES

Next we wish to elucidate the connections between S-group planes and S-planes.

Let (G, S) be any S-group and $E(G, S)$ its group plane; then we consider the map

$$\Phi : \begin{cases} S \to S^* \\ x \mapsto x^* \end{cases}$$

We sometimes denote the map Φ more precisely by $\Phi(G, S)$. It is a mapping of the set S of lines of the group plane $E(G, S)$ into the set of axial collineations of $E(G, S)$, which associates to every line x a collineation with axis x. For its injectivity compare Corollary 4.8. The triplet $(S, \kappa(G, S), \Phi(G, S))$ is not an S-plane, in general. Even if the group plane $E(G, S)$ is Δ-connected, there are nontrivial counter examples—compare Chapter 6, Section 6. But conversely, we may consider every S-plane as the group plane of a suitable S-group:

THEOREM 4.9. *Let* (L, κ, Φ) *be an S-plane containing a triangle and* B *the group generated by* $S := Im \, \Phi$. *Furthermore, we put*

$$(\bar{B}, \bar{S}) := \begin{cases} (B, S) & if \quad 1 \notin S \\ (B \otimes Z_2, S \times \{\bar{z}\}) & if \quad 1 \in S \end{cases}$$

$[Z_2 = \{1, \bar{z}\}$ *denotes the cyclic group of order* 2]. *Then* $(\bar{B}, \bar{S})$ *is an S-group, whose group plane* $(\bar{S}, \kappa(\bar{B}, \bar{S}))$ *is isomorphic to* (L, κ).

PROOF. We use the abbreviations introduced in Chapter 3, Section 4, and denote by J the set of involutions in B, and by $\bar{J}$ the set of involutions in $\bar{B}$. We show that $\bar{S}$ is a set of involutions in $\bar{B}$. This is clear in the case that $1 \notin S$. In the case that $1 \in S$, the elements of $\bar{S}$ are given by $(\sigma_a, \bar{z})$ with $\sigma_a \in S$. Since $(\sigma_a, \bar{z})^2 = (1, 1)$ and $(\sigma_a, \bar{z}) \neq (1, 1)$ (as $1 \neq \bar{z}$), we have $(\sigma_a, \bar{z}) \in \bar{J}$. Next we show that $\bar{S}$ generates $\bar{B}$. Obviously, it suffices to elaborate explicitly the case $1 \in S$. For that let $\bar{\alpha}$ be any element of $\bar{B}$. Then $\bar{\alpha} = (\alpha, 1)$ or $\bar{\alpha} = (\alpha, \bar{z})$, with $\alpha \in B$. Since S generates B, we can find elements $\sigma_1, \ldots, \sigma_n$ in S such that $\alpha = \sigma_1 \cdots \sigma_n$. Then we have

$$(\alpha, 1) = \begin{cases} (\sigma_1, \bar{z}) \cdots (\sigma_n, \bar{z}) & \text{if } n \text{ is even} \\ (1, \bar{z})(\sigma_1, \bar{z}) \cdots (\sigma_n, \bar{z}) & \text{if } n \text{ is odd} \end{cases}$$

and

$$(\alpha, \bar{z}) = \begin{cases} (1, \bar{z})(\sigma_1, \bar{z}) \cdots (\sigma_n, \bar{z}) & \text{if } n \text{ is even} \\ (\sigma_1, \bar{z}) \cdots (\sigma_n, \bar{z}) & \text{if } n \text{ is odd} \end{cases}$$

Since $1 \in S$, we see that all the elements on the right-hand side are elements of $\bar{S}$.

We now verify Axiom S for the pair $(\bar{B}, \bar{S})$.

(a) If $1 \notin S$, then $\bar{S} = S$, $\bar{B} = B$, and $\bar{J} = J$. Let $\sigma_a \neq \sigma_b$, and suppose $\sigma_a \sigma_b \sigma_x$, $\sigma_a \sigma_b \sigma_y$, $\sigma_a \sigma_b \sigma_z \in J$, where σ_a, σ_b, σ_x, σ_y, and σ_z are elements of S.

Hence by [S'] we know that $\sigma_a \neq \sigma_b$ and $\sigma_a\sigma_b\sigma_x$, $\sigma_a\sigma_b\sigma_y$, $\sigma_a\sigma_b\sigma_z \in S \Rightarrow$ (by [S]) $a \neq b$ and (a, b, x), (a, b, y), $(a, b, z) \in \kappa \Rightarrow x, y, z \in L(a, b) \Rightarrow$ $(x, y, z) \in \kappa \Rightarrow$ (by [S]) $\sigma_x\sigma_y\sigma_z \in S$.

(b) Now suppose $1 \in S$. Choose $(\sigma_a, \bar{z}) \neq (\sigma_b, \bar{z})$, and suppose $(\sigma_a, \bar{z})(\sigma_b, \bar{z})(\sigma_x, \bar{z})$, $(\sigma_a, \bar{z})(\sigma_b, \bar{z})(\sigma_y, \bar{z})$, $(\sigma_a, \bar{z})(\sigma_b, \bar{z})(\sigma_z, \bar{z}) \in \bar{J}$, where σ_a, σ_b, σ_x, σ_y, and σ_z are in S. This implies $\sigma_a \neq \sigma_b$ and $(\sigma_a\sigma_b\sigma_x, \bar{z})$, $(\sigma_a\sigma_b\sigma_y, \bar{z})$, $(\sigma_a\sigma_b\sigma_z, \bar{z}) \in \bar{J} \Rightarrow \sigma_a \neq \sigma_b$ and $\sigma_a\sigma_b\sigma_x$, $\sigma_a\sigma_b\sigma_y$, $\sigma_a\sigma_b\sigma_z \in J \cup \{1\} \Rightarrow a \neq b$ and $\sigma_a\sigma_b\sigma_x$, $\sigma_a\sigma_b\sigma_y$, $\sigma_a\sigma_b\sigma_z \in S$ (by [S']) $\Rightarrow a \neq b$ and (a, b, x), (a, b, y), $(a, b, z) \in \kappa$ (by [S]) $\Rightarrow x$, y, $z \in L(a, b) \Rightarrow (x, y, z) \in \kappa \Rightarrow \sigma_x\sigma_y\sigma_z \in S$ (by [S]) $\Rightarrow (\sigma_x\sigma_y\sigma_z, \bar{z}) \in \bar{S} \Rightarrow (\sigma_x, \bar{z})(\sigma_y, \bar{z})(\sigma_z, \bar{z}) \in \bar{S}$. This completes the first part of the proof of the theorem.

To prove the second half, we consider the map

$$\rho : \begin{cases} L \to \bar{S} \\ x \mapsto x\rho := \begin{cases} \sigma_x & \text{if } 1 \notin S \\ (\sigma_x, \bar{z}) & \text{if } 1 \in S \end{cases} \end{cases}$$

From the definition of $\bar{S}$ we know that ρ is surjective. It is injective: Let $x, y \in L$ and $x \neq y$ and $\sigma_x = \sigma_y$. We choose a line z not in $L(x, y)$ and deduce from [S] $\sigma_x\sigma_y\sigma_z = \sigma_x\sigma_x\sigma_z \in S$, since $(x, x, z) \in \kappa$. This implies by [S] $(x, y, z) \in \kappa$ and therefore $z \in L(x, y)$, a contradiction. Thus we have $\sigma_x \neq \sigma_y$ and therefore $x\rho \neq y\rho$. To prove that ρ is an isomorphism we must show that ρ preserves concurrency. We conclude successively that

$$(a, b, c) \in \kappa \overset{[S]}{\Leftrightarrow} \sigma_a\sigma_b\sigma_c \in S \overset{[S']}{\Leftrightarrow} \begin{cases} \sigma_a\sigma_b\sigma_c \in J & \text{if } 1 \notin S \\ (\sigma_a, \bar{z})(\sigma_b, \bar{z})(\sigma_c, \bar{z}) \in \bar{J} & \text{if } 1 \in S \end{cases}$$

$$\Leftrightarrow (a\rho)(b\rho)(c\rho) \quad \text{involutory} \Leftrightarrow (a\rho, b\rho, c\rho) \in \bar{\kappa}$$

This completes the proof of the theorem. ∎

If (L, κ, Φ) is an S-plane, we say the S-group $(\bar{B}, \bar{S})$ in Theorem 4.9 is the S-group *associated* to the S-plane, and we call the isomorphism ρ of (L, κ) onto $(\bar{S}, \bar{\kappa})$ the *canonical isomorphism*.

If $(L(V, Q), \kappa(V, Q), \Phi(V, Q))$ is the metric plane over the metric vector space (V, Q) satisfying [E] and dim $V^{\perp} \leq 1$, then, by Proposition 3.5, $(L(V, Q), \kappa(V, Q), \Phi(V, Q))$ is a Δ-connected S-plane. The S-group associated to this S-plane is denoted by $(\bar{O}(V, Q), \bar{S}(V, Q))$.

We note that Theorem 4.9 is weaker than it seems to be. In spite of the fact that (L, κ) is isomorphic to the group-plane $(\bar{S}, \kappa(\bar{B}, \bar{S}))$, the triplet $(\bar{S}, \kappa(\bar{B}, \bar{S}), \Phi(\bar{B}, \bar{S}))$ need not be an S-plane, hence (L, κ, Φ) is not isomorphic to $(\bar{S}, \kappa(\bar{B}, \bar{S}), \Phi(\bar{B}, \bar{S}))$ in general. But we can prove the following

COROLLARY 4.10. *Let* (L, κ, Φ) *be a complete metric plane, and* $S := Im\ \Phi$ *and* $(\bar{B}, \bar{S})$ *the S-group associated to* (L, κ, Φ). *Then*

$(\bar{S}, \kappa(\bar{B}, \bar{S}), \Phi(\bar{B}, \bar{S}))$ *is an S-plane, and the canonical isomorphism* ρ *of* (L, κ) *onto* $(\bar{S}, \kappa(\bar{B}, \bar{S}))$ *is an isomorphism of* (L, κ, Φ) *onto* $(\bar{S}, \kappa(\bar{B}, \bar{S}), \Phi(\bar{B}, \bar{S}))$. *Moreover, the map* $\bar{\rho} := \alpha \mapsto \rho^{-1}\alpha\rho$ *is an isomorphism of the group* B *generated by* S *onto the group generated by* $\bar{S}^*$.

PROOF. In the notations used above and with $\sigma_a := a\Phi$ for $a \in L$ and $\sigma_{\bar{a}} := \bar{a}\psi$ for $\bar{a} \in \bar{S}$ and $\psi := \Phi(\bar{B}, \bar{S})$, we have, by Lemma 3.2(d), for all $\bar{x}$ in $\bar{S}$ $(x := \bar{x}\rho^{-1})$:

$$\bar{x}\rho^{-1}\sigma_a\rho = (x\sigma_a)\rho = \begin{cases} \sigma_{x\sigma_a} & \text{if } 1 \notin S \\ (\sigma_{x\sigma_a}, \bar{z}) & \text{if } 1 \in S \end{cases}$$

$$= \begin{cases} \sigma_a\sigma_x\sigma_a \\ (\sigma_a\sigma_x\sigma_a, \bar{z}) \end{cases}$$

$$= \begin{cases} (\sigma_x)\sigma_a{}^* \\ (\sigma_x, \bar{z})(\sigma_a, \bar{z})^* \end{cases}$$

$$= (x\rho)(a\rho)^* = \bar{x}((a\rho)\psi) = \bar{x}\sigma_{a\rho}$$

Hence $\rho^{-1}\sigma_a\rho = \sigma_{a\rho}$ for all a in L, and $(\bar{S}, \kappa(\bar{B}, \bar{S}), \Phi(\bar{B}, \bar{S}))$ is an S-plane, isomorphic to (L, κ, Φ). This proves the first statement of Corollary 4.10, and the second is clear. ∎

By Theorem 4.9 and Corollary 4.10 the theory of complete metric planes can be developed in the context of the theory of S-groups with only Δ-connected points, which is treated thoroughly in Chapter 6.

Preparatory to deciding whether for a given S-group (G, S) the triplet $(S, \kappa(G, S), \Phi(G, S))$ is an S-plane we prove the following

THEOREM 4.11. *Let* (G, S) *be an S-group whose group-plane contains a quadrilateral. If* $\alpha^2 = 1$ *for all* α *in* $Z(G) \cap S^4$, *then* $(S, \kappa(G, S), \Phi(G, S))$ *is an S-plane.*

PROOF. As mentioned above, $\Phi(G, S)$ is a map of the set S of lines of $E(G, S)$ into the set of all the axial collineations of $E(G, S)$. We have $a^* = a\Phi(G, S)$ and $(a^*)^2 = 1$ for all a in S and $S^* = \text{Im}\,\Phi(G, S)$.

To prove [S] for $(S, \kappa(G, S), \Phi(G, S))$, let $(a, b, c) \in \kappa(G, S)$. Then $abc \in S$, and there is a line d such that $abc = d$. This yields $a^*b^*c^* = d^*$ and therefore $a^*b^*c^* \in S^*$.

Conversely, let $a^*b^*c^* \in S^*$. Then there is a line d such that $a^*b^*c^* = d^*$. This yields $x^{abcd} = x$ for all x in S and therefore $abcd \in Z(G) \cap S^4$. By our assumption we have $(abcd)^2 = 1$. From Theorem 4.4(d) we see that $abcd = 1$, and therefore $abc \in S$ and $(a, b, c) \in \kappa(G, S)$.

To prove [S'], let $a^*b^*c^*$ be involutory. Then $(a^*b^*c^*)^2 = 1$ and

$a^*b^*c^* \neq 1$. This yields $x^{(abc)^2} = x$ for all x in S and therefore $(abc)^2 \in Z(G)$. From $(abc)^2 = b^a c^a bc \in S^4$, the assumption $(abc)^4 = 1$ and Theorem 4.4(d) it follows that $(abc)^2 = 1$ and $abc \neq 1$, hence $abc \in J$, and therefore $abc \in S$ and $a^*b^*c^* \in S^*$, as required.

This proves the theorem. ∎

By Theorem 4.11, if (G, S) is an S-group whose group plane contains a quadrilateral, and which satisfies $\alpha^2 = 1$ for all $\alpha \in Z(G) \cap S^4$, then

$$abc \in S \Leftrightarrow a^*b^*c^* \in S^* \qquad (\circ)$$

This equivalence is not true in an arbitrary S-group (G, S).

5

S-GROUP PLANES WITH Δ-CONNECTED POINTS

The basic concept for the characterization of Euclidean and non-Euclidean planes in the framework of S-group planes is the concept of Δ-connected points. In view of this fact we compile some relevant results on S-group planes with Δ-connected points. First, we show that for an S-group (G, S) whose group plane contains at least one Δ-connected point any element α of G is a product of not more than four elements of S (reduction theorem).

1. THE REDUCTION THEOREM

THEOREM 5.1 (Reduction). *Let* (G, S) *be any S-group whose group plane contains at least one Δ-connected point. Then every element in G is a product of at most four elements in S.*

PROOF.   It is sufficient to prove that $S^5 \subset S^3$. Let A be a Δ-connected point. We show

(1)   *Given* $abc \in S^3$, *there exists some* $a' \in A$ *such that* $a'abc \in S^2$.

We first remark that if $abc \in S$, then trivially $a'abc \in S^2$ for all $a' \in A$. So assume $abc \notin S$. Then $S(ab)$, $S(ac)$, and $S(bc)$ are three distinct points that are pairwise connected by a line. Since A is Δ-connected, there is a line $a' \in A$ such that $a' \in S(ab)$, or $a' \in S(ac)$, or $a' \in S(bc)$. Hence either $(a'ab)c \in S^2$, or $a'abc = (a'ac)b^c \in S^2$, or $a'abc = (a'aa')(a'bc) \in S^2$. This proves (1).

Now let $a_1 a_2 a_3 a_4 a_5 \in S^5$. By repeated use of (1) we can find elements $a_1', a_2', a_3' \in A$ such that $a_1'a_1a_2a_3 \in S^2$, $a_2'(a_1'a_1a_2a_3)a_4 \in S^2$, and $a_3'(a_2'a_1'a_1a_2a_3a_4)a_5 \in S^2$. But $a_1'a_2'a_3' \in A \subset S$, and thus $a_1 a_2 a_3 a_4 a_5 = a_1'a_2'a_3'(a_3'a_2'a_1'(a_1a_2a_3a_4a_5)) \in S^3$. This proves the theorem.   ∎

If the Δ-connected point whose existence is assumed in the theorem happens to be completely connected, we can improve Theorem 5.1 and its proof to obtain:

PROPOSITION 5.2. *Let* (G, S) *be any S-group whose group plane contains at least one completely connected point. Then every element of* G *is a product of at most three elements of* S.

PROOF.   It is sufficient to prove that $S^4 \subset S^2$. Let A be a completely connected point. We show

(1)   *Given* $ab \in S^2$, *there exists* $a' \in A$ *such that* $a'ab \in S$.

If $a = b$, this holds for all $a' \in S$. If $a \neq b$, then the points $A, S(ab)$ are connected by at least one line a', and thus $a' \in A$ and $a'ab \in S$. This proves (1).

The theorem now follows from (1) by an argument similar to the one used in Theorem 5.1. ■

We have:

COROLLARY 5.3. *Let* (G, S) *be an S-group whose group plane contains a completely connected point, and a quadrilateral. Then the center* $Z(G)$ *of* G *contains at most one element* $\neq 1$, *and this element, if it exists, lies in* S.

PROOF.   This follows immediately from Proposition 5.2 and Corollary 4.7 ■

We would like to point out here that the statement of Theorem 5.1 (the reduction theorem) cannot in general be improved. For if we let $S(M)$ be the set of reflections of a Minkowskian coordinate plane M [compare Chapter 3, Section 1(b)], and $B(M)$ be the group of motions of M, then $(B(M), S(M))$ is an S-group whose group plane is Δ-connected; however, there are elements in $B(M)$ that cannot be written as a product of less than four elements of $S(M)$. Compare (1) in the proof of Proposition 6.12.

Also, Proposition 5.2 cannot be improved. To see this we consider the set $S(E)$ of reflections of a Euclidean coordinate plane, together with its group of motions $B(E)$. Again, $(B(E), S(E))$ is an S-group whose group plane contains completely connected points, whereas $B(E)$ contains elements that cannot be expressed as a product of less than three elements of $S(E)$.

2. THE THEOREM OF THE SET OF PERPENDICULARS TO A LINE

In this section we prove that in an S-group plane $E(G, S)$ with Δ-connected points the product of three lines that are orthogonal to a given line x is itself a line, provided that x is not in the center of the group G. Equivalently, in such an S-group the set $x^\perp$ of perpendiculars to a line x not in the center of the group G is contained in a point.

THEOREM 5.4 (Set of perpendiculars). *Let* (G, S) *be any* S-*group whose group plane contains at least one* Δ-*connected point. Then for any* x *not in the center of* G *we have that*

$$u, v, w \in x^\perp \quad implies \quad uvw \in S \qquad (*)$$

PROOF. If $x \in S^2$, then $x^\perp$ is the point $S(x)$. Hence we obtain $uvw \in S$ by Axiom S.

Thus we assume $x \notin S^2$. First, we prove the following

(1) *For all* s, t $\in x^\perp$, *we have* st $\in J \Leftrightarrow x \in S(st)$.

Clearly, $xs = sx$, $xt = tx$, and $xst \neq 1$, as otherwise $x \in S^2$. We therefore deduce that $st \in J \Leftrightarrow st = ts$ and $s \neq t \Leftrightarrow xst \in J$ and $s \neq t \Leftrightarrow x \in S(st)$.

Now, let u, v, w be three lines in $x^\perp$, and assume $uvw \notin S$. Then u, v, w are not concurrent. Define $U := S(vw)$, $V := S(uw)$ and $W := S(uv)$. Clearly, $x \neq u$, v, w, and thus x belongs to at most one of U, V, W.

Suppose $x \in U$. We replace u by $w' := w^u$. Clearly, u, v, $w' \in x^\perp$, and $uvw' = uvuwu = (vuw)^u \notin S$, as otherwise $uvw \in S$. If $x \in S(vw) = U$, and $x \in S(vw')$, then $x \neq v$ implies that $S(vw) = S(vw')$. Hence w, $w' \in S(vw)$, $S(uw)$. Therefore either $w = w'$, and $wu \in J$, which implies $x \in V$ by (1), a contradiction, or $S(uw) = S(vw)$, and $uvw \in S$, which is also contradictory. Thus $x \notin S(vw^u)$, and so we may assume without loss of generality that $x \notin U$, V, W.

This implies that $U \cup V \cup W \subset x^\perp$, and by (1), we know uv, uw, $vw \notin J$. Next, we show

(2) *There exists a* Δ-*connected point* A *not on* x.

By the assumptions of the theorem we know that there exists at least one Δ-connected point B. If $x \notin B$, we let $A := B$. Now suppose $x \in B$. We know B is connected with one of the points U, V, W. Suppose without loss of generality that B is connected with W by a line u'. Let $v' := u'uv$. Then $v' \notin B$ and $B = S(u'x)$. Since $v' \in W \subset x^\perp$, we know that $x \in B, B^{v'}$. If $B = B^{v'}$, then $B \subset (v')^\perp$, and so $u'v' = uv \in J$, a contradiction. Thus $B \neq B^{v'}$.

Since x is not in the center of G, we can find a line y such that $xy \neq yx$. Then $y \notin B$ or $y \notin B^{v'}$.

If $y \notin B$, then $x \notin B^y$, as otherwise $x \in B, B^y$, implying that $xy = yx$. Similarly, if $y \notin B^{v'}$, then $x \notin B^{v'y}$. Hence our assertion (2) holds if we put $A := B^y$ or $A := B^{v'y}$.

Next, we show

(3) *If* A *is a* Δ-*connected point, then* $A^x = A$.

If $x \in A$, there is nothing to prove. So suppose $x \notin A$. Assume first that $u, v, w \notin A$. Without loss of generality, we assume that A is connected to U by a line s. Our choice of A assures us that $s \neq v, w$.

Now, $U \neq U^u$, as otherwise $U \subset u^\perp$, and thus $uv \in J$. If U^u does not lie on s, then A is connected to one of U^u, V, W by a line $t \neq s$, and $t, s \in x^\perp$. Thus $A = S(st)$, which implies that $A^x = S(st)^x = S(st) = A$.

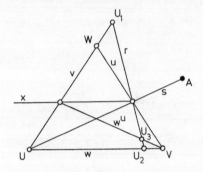

Now suppose U^u lies on s. Then $s \in U, U^u$, hence $s \in u^\perp$, or $us \in J$. But by (1), this implies $x \in S(us)$.

Consider also the point $S(vw^u)$. If $s \in S(vw^u)$, then $w^u \in S(vs) = U$, and $w \in U$, implying $wu \in J$, a contradiction. Hence $s \notin S(vw^u)$. If $x \notin S(vw^u)$, then A is connected to one of $S(vw^u)$, .V, W by a line $t \neq s$, and again $t, s \in x^\perp$, implying $A^x = A$. So we also assume $x \in S(vw^u)$.

Now define $r := xsu$. Clearly, $r \neq x, s, u$.

Observe $rx = su \in J$, and thus $r \in x^\perp$. Let $U_1 := S(rv)$, $U_2 := S(rw)$ and $U_3 := S(rw^u)$. If $U_1 = U_2$, then $r \in S(vw)$; consequently $r = s$; if $U_1 = U_3$, then $r \in S(vw^u)$, consequently $r = x$; if $U_2 = U_3$, then $r \in S(ww^u)$, consequently $r = u$. Hence U_1, U_2, U_3 are mutually distinct points.

Also, $x, s \notin U_i$, $i = 1, 2, 3$. Thus $U_1, U_2, U_3 \subset x^\perp$, and so A is connected to one of U_1, U_2, U_3 by a line $t \neq s$, $t \in x^\perp$. Thus again $A^x = A$.

Now suppose A lies on u, and $x \notin A$. If $A = V$ or $A = W$, then $A \subset x^\perp$, hence $A^x = A$. If $A \neq V, W$, consider A^v. Clearly, $u \notin A^v$, as otherwise $u \in A, A^v$, implying $uv \in J$, a contradiction. Also, $v \notin A^v$, as otherwise $v \in A$, and $A = S(uv) = W$, a contradiction. If $w \in A^v$, then $u, w^v \in A$, and thus $A = S(uw^v)$, implying $A^x = A$.

If $u, v, w \notin A^v$, then by the above $A^{vx} = A^v$. But $vx = xv$, and so $A^v = A^{vx} = A^{xv}$, and consequently $A = A^x$.

This proves (3).

Now we prove

(4) $y \in x^\perp$ *for all lines* $y \neq x$.

Choose any line $y \neq x$. If $y \in U$ or $y \in V$ or $y \in W$, then $y \in x^\perp$, hence there is nothing to prove. Now let $y \notin U, V, W$.

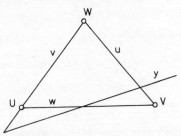

From (2) we know that there exists a Δ-connected point A not on x.

If $u, v, w \notin A$, then A is connected to one of $S(yu)$, $S(yv)$, $S(yw)$ by some line a'. Without loss of generality, suppose A and $S(yu)$ are connected by $a' \neq u$. Then $S(yu) = S(a'u)$. If $x \notin S(yu)$, then $S(yu)^x = S(a'u)^x = S(a'u) = S(yu)$, hence $y \in S(yu) \subset x^\perp$.

If $x \in S(yu)$, we have the following situation:

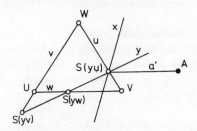

If W is connected to $S(yw)$ by some line $w' \in W$, then $S(yw) = S(ww')$, and $x \notin S(yw) \Rightarrow S(yw)^x = S(ww')^x = S(ww') = S(yw) \Rightarrow y \in x^\perp$.

If W is not connected to $S(yw)$, then v^u is a line through W, $v^u \neq v, u$, and $S(yv^u) \neq S(yv), S(yw)$. Also, $x \notin S(yv), S(yw), S(yv^u)$, and so A is connected to one of $S(yv), S(yw), S(yv^u)$, and we are again finished provided that $v^u \notin A$. If $v^u \in A$ and $A \cap S(yv) = \varnothing = A \cap S(yw)$, then $w^u \notin A$ [as otherwise $A = S(vw)^u = U^u$ and $w \in A$]. Furthermore, we have $S(yw^u) \neq S(yw)$ and $x \notin S(yw^u)$. If $w^u \in S(yv)$, then $S(yv) = S(vw^u) \subset x^\perp$ and $y \in x^\perp$, as required. If $w^u \notin S(yv)$, then $S(yw^u) \neq S(yv)$ and therefore $A \cap S(yw^u) \neq \varnothing$. The line connecting A and $S(yw^u)$ is distinct from w^u, and therefore we obtain $S(yw^u) \subset x^\perp$ and $y \in x^\perp$.

Finally, suppose the Δ-connected point A lies on one of the lines u, v, w, say u. If $A = W$, then $A^w = W^w$ is a Δ-connected point and $u, v, w \notin A^w$, so that we may proceed as above. The same argument holds if $A = V$.

If $u \in A$, and $A \neq V$, W, then again consider A^w. If $u \in A^w \Rightarrow u^w \in A \Rightarrow A = S(uu^w) = V$, a contradiction. If $w \in A^w \Rightarrow w \in A \Rightarrow A = V$; $u, w \notin A^w$. If $v \notin A^w$, we proceed as above.

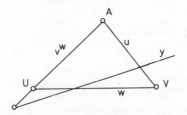

Otherwise, we consider the triangle U, V, A and the Δ-connected point A^w with $x \notin A^w$. Then we see that $u, w \notin A^w$, and also $v^w \notin A^w$. From here we proceed as above.

Thus we have shown that the assumption that

$$uvw \notin S$$

leads to the following result: $xy = yx$ for all $y \in S$, or $x \in Z(G)$, a contradiction. ∎

If the Δ-connected point whose existence is postulated in Theorem 5.4 happens to be completely connected, or if all the points are Δ-connected, then the proof of Theorem 5.4 can be shortened considerably. As both of these cases are of particular interest, we reprove Theorem 5.4 under these assumptions.

PROPOSITION 5.5. *Let* (G, S) *be an S-group whose group plane contains at least one completely connected point. Then* (∗) *of Theorem* 5.4 *holds for any* x *not contained in the center of the group* G.

PROOF. To prove (1) and (2) we proceed as in the proof of Theorem 5.4. The proof of (3) becomes much simpler.

(3) $B^x = B$ *for any completely connected point.*

For B is connected with each of the points U, V, W by lines u', v', w', respectively. Since U, V, W are not collinear, at least two of these lines are distinct. Say $u' \neq v'$. But then $B^x = S(u'v')^x = S(u'^x v'^x) = S(u'v') = B$.

Since x is not in the center of G, there exists a line y such that $xy \neq yx$. By (2) there also exists a completely connected point A such that $x \notin A$.

By (3) $A^x = A$, and this yields $A \subset x^\perp$, hence $y \notin A$. Since A is completely connected, there exist lines $s, t \in A$ such that $s \in A^y$ and $t \in S(xy)$.

We have $sy \in J$. If also $st \in J$, then by (1) $x \in S(st) = A$, contradicting the choice of A. Hence $st \notin J$. Now suppose $sxy \notin J$. Since $x, y \in s^\perp$, we deduce $S(xy) \subset s^\perp$, hence in particular $t \in S(xy) \subset s^\perp$, implying $st \in J$, a contradiction. Thus $sxy \in J$. But $sx = xs$, $sy = ys$, and $sxy = yxs$ allows us to conclude $xy = yx$, a contradiction.

Thus the assumption that $uvw \notin S$ is false, and Proposition 5.5 is proved. ■

PROPOSITION 5.6. *Let* $E(G, S)$ *be an S-group plane whose points are all* Δ*-connected. Then* $(*)$ *of Theorem 5.4 holds for all lines* x *not in the center of the group* G.

PROOF. As in the proof of Theorem 5.4 we may assume that $x \notin S^2$ and that there exist three lines $u, v, w \in x^\perp$ such that $uvw \notin S$. Again we put $U := S(vw)$, $V := S(uw)$ and $W := S(uv)$ and, as in Theorem 5.4, we can assume that $x \notin U, V, W$.

By assumption, x is not in the center of the group G, hence we can find a line y such that $xy \neq yx$. Since $uvw \notin S$ and $x \notin U, V, W$, at most one of the lines u, v, w lies in $S(xy)$, and we may assume $u \notin S(xy)$. Let $X := S(uy)$. Clearly, $X^x \neq X$, as otherwise we would have $y^x, y \in S(xy)$, X, and $S(xy) \neq X$, which would imply $y^x = y$, or $xy = yx$, in contradiction to the choice of y.

The point U is not connected with X, for suppose there is a line $z \in U, X$. Then $z \neq u$, and thus $X = S(uz)$, which implies $X^x = S(uz)^x = S(u^x z^x) = S(uz) = X$, a contradiction. Similarly, U is not connected with X^x, as otherwise $U^x = U$ is connected with $(X^x)^x = X$, a contradiction. We indicate the situation in the following diagram:

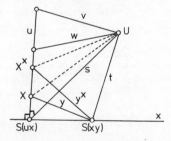

Since $y^x \in X^x$, $S(xy)$, and since U is Δ-connected, there exist lines $s, t \in U$ such that $s \in S(ux)$, and $t \in S(xy)$.

Then we have $ux = xu$, $sx = xs$ and $uxs = sxu$, and so we conclude $su = us$. Thus X^s is a point on u, and since U, X are not connected, also

$U^s = U, X^s$ are not connected. Since U is Δ-connected, it is not connected to at most two points on u, namely to X and X^x. Hence we conclude that either $X^s = X$ or $X^s = X^x$. If $X^s = X$, we obtain $sy \in J$, and so $S(xy)^s = S(x)$. Also, $s \notin S(xy)$, as otherwise $x, s \in S(xy), S(ux)$, and $x \neq s$ imply that $S(xy) = S(ux)$, and so $u \in S(xy)$, a contradiction. Thus we may conclude $S(xy) \subset s^{\perp}$, and thus in particular, $st \in J$. By (1) in the proof of Theorem 5.4, we then obtain $x \in S(st) = U$, a contradiction.

Thus we are left with the second alternative: $X^s = X^x$. Let $r := uxs$. Then $r \in x^{\perp}$ and $X^r = X^{uxs} = X$. If $r = t$, then $ux = st \in J$, which implies $x \in S(st) = U$, a contradiction. Hence $r \neq t$. An argument similar to the one above with r instead of s now shows that $x \in S(rt)$. But then $r, x \in S(rt)$, $S(ux)$, and $r \neq x$ implies $S(rt) = S(ux)$. Thus $t, s \in S(ux)$, U and $t \neq s$; hence $S(ux) = U$ and $x \in U$, a contradiction.

Thus the assumption $uvw \notin S$ is false, proving Proposition 5.6. ∎

3. THE MAPPING σ_{qAB} (GEGENPUNKTPAARUNG)

We now introduce mappings that are of great importance in the investigation of S-groups with Δ-connected points.

Let (G, S) be an arbitrary S-group, and let A, B be points of its group plane which are not connected by a line. Let q be any line not contained in A. We then claim: For any line $x \in A$ there is at most one line b in B such that $qxb \in S$. For suppose $qxb, qxb' \in S$ for $b, b' \in B$. Then $b, b' \in B, S(qx)$. Now $S(qx)$ is defined, as $q \neq x$ since $q \notin A$ and $x \in A$. Also, $B \neq S(qx)$ since $x \in A$, and thus $x \notin B$. Hence $b = b'$. Thus

$$\sigma : x \mapsto qxb \quad \text{for} \quad qxb \in (qxB) \cap S$$

is a mapping from a subset of A into the complex $qAB \cap S$ in the group G. When necessary, we denote σ more precisely by σ_{qAB}.

Obviously, we have: $x\sigma$ exists $\Leftrightarrow qxB \cap S \neq \varnothing \Leftrightarrow B, S(qx)$ are connected by a line.

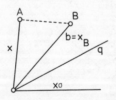

If $B, S(qx)$ are connected by a line, we then denote this joining line by x_B. We now let σ be a mapping σ_{qAB}. We prove in succession:

(σ1) σ is injective.

$x\sigma = y\sigma$ implies $qxx_B = qyy_B$, and so $xy = x_By_B$. If $x \neq y$, then $A = S(xy) = S(x_By_B) = B$, contradicting the fact that A, B are not connected by a line. Thus $x = y$. ∎

$(\sigma 2)$ *If* $x \neq y$, *and* $x, y \in A$, *then* B *and* $S(x\sigma y\sigma)$ *are not connected.*

Suppose $v \in B, S(x\sigma y\sigma)$. Then $(xy)(y_Bvx_B) = (x\sigma y\sigma v)^{x_B} \in J$, hence $y_Bvx_B \in S(xy) = A$. But $y_Bvx_B \in B$, which implies that A, B are connected, a contradiction. ∎

$(\sigma 3)$ *If* B *is* Δ-*connected, then* $x, y, z \in A$ *implies* $x\sigma y\sigma z\sigma \in J$.

Suppose $x\sigma y\sigma z\sigma \notin J$. Then the Δ-connected point B is connected with one of the three points $S(x\sigma y\sigma)$, $S(x\sigma z\sigma)$, or $S(y\sigma z\sigma)$, contradicting $(\sigma 2)$. ∎

$(\sigma 4)$ *If* B *is* Δ-*connected, then at most one line of* A *is not in the domain of* σ.

Let x, x' be distinct lines of A. Then B is connected with at least one of $S(xq)$, $S(x'q)$ since B is not connected with A. Thus one of x, x' is in the domain of σ. ∎

$(\sigma 5)$ *If* B *is* Δ-*connected, and if* A *contains at least three lines, then there is exactly one point* C *such that* $\operatorname{Im} \sigma \subset C$.

This is clear by $(\sigma 1)$ to $(\sigma 4)$. ∎

$(\sigma 6)$ *If* A *is* Δ-*connected, then for* $x, y, z \in A$, *we have* $x\sigma y\sigma z\sigma \in S$.

If $q \in B$ then $q \in B$, $S(qx)$ and therefore $x\sigma = qxx_B = qxq = x^q$ for all x in A. This implies $x\sigma y\sigma z\sigma = x^q y^q z^q = (xyz)^q \in J$ for all x, y, z in A, as required.

Now let $q \notin B$. Then the mapping $\tau := \sigma_{qBA}$ is defined. If $x\sigma$ is defined for $x \in A$, then

$$x\sigma = qxx_B = (qx_Bx)^q = x_B\tau$$

Thus, if $x, y, z \in A$ and x, y, z are in the domain of σ, then x_B, y_B, z_B exist and are contained in the domain of τ. We have $x\sigma y\sigma z\sigma = (x_B\tau y_B\tau z_B\tau)^q \in J$ since $x_B\tau y_B\tau z_B\tau \in J$ by $(\sigma 3)$. This proves $(\sigma 6)$. ∎

$(\sigma 7)$ *If* A *is* Δ-*connected, and if* B *contains at least three lines, then there are at least two lines in the domain of* σ.

If $q \in B$, then $A^\sigma = A^q$ and $\operatorname{dom} \sigma = A$, and our assertion is true. Now let $q \notin B$, and let b, b', b'' be three distinct lines in B. Since A is Δ-connected, but A is not connected with B, it follows that A is connected with at least two of the points $S(bq)$, $S(b'q)$ or $S(b''q)$ by lines x, y. Clearly, x, y are in

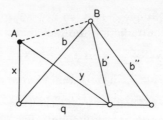

the domain of σ, and $x \neq y$, as otherwise $b, b' \in B, S(qx)$, and $b \neq b'$, which implies that $B = S(qx)$ and $x \in A, B$, a contradiction. ∎

(σ8) *If A is Δ-connected, and if B contains at least three lines, then there exists exactly one point C such that Im $\sigma \subset$ C.*

This follows at once from (σ1), (σ6), and (σ7). ∎

Now let σ be the mapping σ_{qAB}. If there exists exactly one point C such that Im $\sigma \subset C$, then we denote this point by A^σ, and we say "the point A^σ exists."

We collect our results in the following:

THEOREM 5.7 (The mapping σ_{qAB}). *In an S-group plane let A, B be two points not connected by a line. Let q be any line such that $q \notin A$, and define $\sigma := \sigma_{qAB}$. If one of the points A, B is 2-Δ-connected, or if one of the points A, B is Δ-connected, and the other contains at least three lines, then the point A^σ exists. Moreover, σ is injective, and the points A^σ, B are not connected by a line.*

Let A, B be two points in an incidence structure. We call the pair A, B a Δ-*pair* either if the two points are not connected but one of them is 2-Δ-connected, or if one of them is Δ-connected and the other contains at least three lines. Thus we see that Theorem 5.7 deals with Δ-pairs.

We now introduce the following notation: Let A, B be two points of an S-group plane which are not connected by a line. We then denote by the symbol $[A, B]$ the set of all points X for which there exists a line q such that the mapping $\sigma := \sigma_{qBA}$ is defined, and such that the point B^σ exists and coincides with X. In short, we may write

$$[A, B] := \{B^\sigma \mid \sigma := \sigma_{qBA} \quad \text{for a suitable} \quad q\}$$

We now prove:

LEMMA 5.8. *Let A, B be any Δ-pair in an S-group-plane, and let a be a line not in A. Then there exists a point C such that $a \in C \in [A, B]$.*

PROOF. From the proofs of $(\sigma 4)$ and $(\sigma 7)$ we see that there exist lines b, c such that $b \in A$, $c \in B$, and $abc \in S$. We put $q := abc$. Since $a \notin A$, we obviously deduce $a \neq b$. Moreover, we have $q \notin B$. For $q \in B$ would imply that $q, c \in B$, $S(ab)$, and, since $B \neq S(ab)$ (as $b \notin B$), we could deduce that $q = c$, hence $a = b$, a contradiction. Thus we see that the mapping $\sigma := \sigma_{qBA}$ is defined, and so Theorem 5.7 yields the existence of B^{σ}. But $c\sigma = qcb = a$, and we conclude that $a \in B^{\sigma}$. Thus our assertion holds if we put $C := B^{\sigma}$. ∎

We now have:

COROLLARY 5.9. *Let A be a Δ-connected point in an S-group plane. Let b be any line not through A. If A is not completely connected, there exists a point B on b which is not connected with A.*

PROOF. Since A is not completely connected, there exists some point C not connected with A. Then either A, C is a Δ-pair, or $|C| = 2$. If A, C is a Δ-pair, the assertion follows at once from Theorem 5.8. In the latter case we may assume that all points not connected with A contain precisely two lines. If $b \in C$, then we may put $B := C$. Now let $b \notin C$ and $C = \{c, c'\}$. If $A \cap S(bc) = \varnothing$ or $A \cap S(bc') = \varnothing$, then our assertion holds. Hence we may assume that there exist lines $a, a' \in A$ such that $abc, a'bc' \in S$. We now put $q := bac$ and $q' := ba'c'$. Then $q, q' \notin C$, and both $\sigma := \sigma_{qCA}$ and $\sigma' := \sigma_{q'CA}$ are defined. If $|\text{dom } \sigma| = 2$ or $|\text{dom } \sigma'| = 2$, we may put $B := C^{\sigma}$, or $B := C^{\sigma'}$, and we are done.* So we are left with the case that $|\text{dom } \sigma| = 1 = |\text{dom } \sigma'|$. In this case A is not connected with either $S(qc')$ or $S(q'c)$, which implies that $|S(qc')| = 2 = |S(q'c)|$. Hence $qc' = c'q$, implying $S(ab)^{c'} = S(cq)^{c'} = S(cq) = S(ab)$. Since $c' \notin S(ab)$, we conclude $S(ab) \subset c'^{\perp}$, hence ac', $bc' \in J$. Repeating the argument with $q'c$, we also obtain $a'c$, $bc \in J$. Hence $bcc' = acc' = 1$ by ① and since bcc', $acc' \notin J$. This implies $a = b$, a contradiction. Hence this last case cannot occur. This proves our assertion. ∎

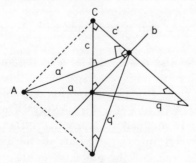

* Here dom σ denotes the domain of the map σ.

4. DROPPING A PERPENDICULAR TO A GIVEN LINE THROUGH A GIVEN POINT

For an incidence structure with orthogonality we can ask whether there exists a perpendicular through a point A to a given line b, that is, whether the set $A \cap b^\perp$ is not the empty set. Obviously, we cannot prove $A \cap b^\perp \neq \varnothing$ for all choices of A and b in an S-group plane. Moreover, even if the S-group plane contains Δ-connected points, it can happen that $A \cap b^\perp = \varnothing$ for some A and b. However, we can show that this has consequences that are important for the further investigation. These consequences can be proved by applying the mapping σ_{qAB}.

LEMMA 5.10. *Let* A *be a* Δ-*connected point in an S-group plane. Let* b *be any line not through* A. *If* $A \cap b^\perp = \varnothing$, *then there exists a point* B *on* b *not connected with* A *such that* $B^{A^2} = B$.

If B *is* Δ-*connected, it is in fact* 2-Δ-*connected.*

PROOF. We let a be any line of A, and define $q := b^a$. From $A \cap b^\perp = \varnothing$ and $b \notin A$, we obtain $A \cap A^b = \varnothing$, hence $A \cap A^q = \varnothing$. Obviously, $q \notin A$ and therefore $q \notin A^q$. Hence the mapping $\sigma := \sigma_{qA^qA}$ is defined. Also, $x \in A \cap S(qx^q)$ for all $x^q \in A^q$, and thus dom $\sigma = A^q$. Hence Theorem 5.7 tells us that the point $B := (A^q)^\sigma$ exists and is not connected with A. From

$$(x^q)\sigma = qx^qx = q^x = b^{ax} \quad \text{for all} \quad x \in A$$

we deduce that $b^{A^2} \subset (A^q)^\sigma = B$. Clearly, the map $\delta : \alpha \mapsto b^\alpha$ for $\alpha \in A^2$ is injective, because $b^\alpha = b$ for some $\alpha \neq 1$ in A^2 implies that $\alpha^b = \alpha$, or $A^b = A$, and since $b \notin A$, we would therefore know that $A \subset b^\perp$, a contradiction. Hence, for any $\beta \neq 1$, $\beta \in A^2$, we have $b \neq b^\beta$, and so $B^\alpha = S(bb^\beta)^\alpha = S(b^\alpha b^{\beta\alpha}) = B$ for all $\alpha \in A^2$. This yields $B^{A^2} = B$.

Next suppose B is Δ-connected. We wish to show that B is 2-Δ-connected. As a first step, we show

(1) *Suppose* C *is a point that is connected with* A *but not with* B. *Then* C = A.

To prove (1), we suppose $A \neq C$, and let a be the line joining A and C. Let $x, y \neq a$ be two lines through A, respectively, through C. Then $x \neq y$, and $B^a \cap S(xy) \neq \varnothing$, since B is by assumption Δ-connected, and B is not connected with either A or C. Let $q = B^a \cap S(xy)$. Then $q \notin C$, and so the mapping $\sigma := \sigma_{qCB}$ is defined. Clearly, $q^a \in B \cap S(qa)$; also, $q^x \in B \cap S(qy)$, since $B^{ax} = B$ by the first part of the proof. Hence $a, y \in$ dom σ, and so by Theorem 5.7 the point C^σ exists and is not connected with B. From $a\sigma = qaq^a = a$, we see that $C^\sigma = A$ or $C^\sigma = C$. If $C^\sigma = A$,

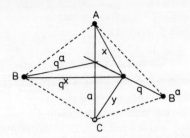

then $x, y\sigma \in A$, $S(qy)$ and $A \neq S(qy)$ implies $x = y\sigma$. But this implies that $x = qyq^x = qyxqx$, or $qyxq = 1$, or $yx = 1$, hence $x = y$, a contradiction. Thus we are left with $C^\sigma = C$, which means, in particular, that $y\sigma = y$. But this yields $xy = q \cdot qxy = q \cdot yxq = qyxqx \cdot x = (y\sigma)x = yx$. If $axy = 1$, then $b, b^y = b^{ax} \in B, S(by)$, and $B \neq S(by)$, which implies that $b = b^y$, and therefore $b \in y^\perp = S(ax) = A$, contradicting the choice of A, b. Hence we have $xy \in J$, and $axy \neq 1$ for any pair of lines $x, y \neq a$ in A, respectively C. Clearly, $|A| = |C| = 2$ is impossible, as otherwise $axy = 1$, which we already led to a contradiction. Similarly, $|A|, |C| > 2$ is impossible, as in this case $A \subset y^\perp$ for all $y \in C$ and $y \neq a$, and $C \subset x^\perp$ for all $x \in A$ and $x \neq a$, again implying $axy = 1$. We are left with the case $|A| = 2$ and $|C| > 2$, or $|A| > 2$ and $|C| = 2$. If $|A| = 2$ and $|C| > 2$, then no two lines of C are orthogonal. Hence $A^y \neq A$ for $y \in C$, $y \neq a$, and $x \in A^y$ for $x \neq a$ in A. Then A^y, B are connected by a line b' and, since $b' \neq x$, a^y, we have $2 > |A^y| = |A| = 2$, a contradiction. Similarly, the case $|A| > 2$ and $|C| = 2$ leads to a contradiction. Hence, the assumption $C \neq A$ is false, and (1) is proved.

Next we show

(2) B *is not* 1-Δ-*connected*.

Again, we assume the contrary. Then there exist distinct connected points C and D, neither of which is connected to B. From (1) we deduce that A is not connected with either C or D. Let c be the line connecting C and D. Let a be any line in A. Then $a \neq c$, and by (1) we know that $B \cap S(ac) \neq \varnothing$. Let $d = B \cap S(ac)$, and define $q := adc$. Then $q \notin A$, and so the mapping $\sigma := \sigma_{qAB}$ is defined. From (1) we deduce that $B \cap S(qx) \neq \varnothing$ for all $x \in A$, hence dom $\sigma = A$. Then Theorem 5.7 yields the existence of the point A^σ which is not connected with B. Since $a\sigma = qad = c$, we see that A^σ lies on c. So we may assume without loss of generality that $A^\sigma = C$. Also, we know $q \notin D$, and so the map $\tau := \sigma_{qDB}$ is defined. Suppose $|$dom $\tau| \geq 2$. Then from Theorem 5.7 we again deduce the existence of the point D^τ, which is not connected with B, and which lies on a, since $c\tau = qcd = a$. Thus by (1) we see that $D^\tau = A$. But for any line $y \neq c$ in dom τ, we obtain $y\tau \in A =$ dom σ, and thus $y\tau\sigma \in C$. Also

$y = q(qyy_B)y_B = y\tau\sigma \in C$, D implies $y = c$, a contradiction.* Hence we conclude that $|\text{dom } \tau| = 1$, and $|D| = 2$.

Let e be the line $\neq c$ in D and $E := S(eq)$. From $|\text{dom } \tau| = 1$ we see that $B \cap E = \emptyset$. By (1) we have $A \cap E = \emptyset$. Thus E and D are the only two distinct points on e not connected with A and B. We choose any line $a' \neq a$ in A. By (1) there exists a line d' in $B \cap S(ca')$. We put $q' := a'd'c$ and $\tau' := \sigma_{q'DB}$. Then we can assume $|\text{dom } \tau'| = 1$, as otherwise by our arguments given above we are led to a contradiction. Hence $B \cap S(eq') = \emptyset$ and therefore $S(eq') = E$, since $S(eq') \neq D$. Next we wish to prove that $E^c = E$. If $E^c \neq E$, then $A \cap E^c \neq \emptyset$, since $E^c \neq D$. Without loss of generality we may assume that $a \in A, E^c$. Then we have $a^c, q \in E, S(ac)$, and $E \neq S(ac)$ (since $a \in A$ and $A \cap E = \emptyset$). This implies $a^c = q$. Since $q = cda$, the equation $a^c q = 1$ is equivalent to $cd^a = 1$. By the first part of the proof we have $B^{aa'} = B$ and therefore $c^{a'} = d^{aa'}$, $d' \in B$, $S(ca')$, and $B \neq S(ca')$ (since $c \notin B$). This yields $c^{a'} = d'$ and thus $cd'^{a'} = 1$. The last equation is equivalent to $a'^c q' = 1$, since $q' = cd'a'$. Therefore we have $a, a' \in A$, E^c and $a \neq a'$, hence $A = E^c$ and $e \in A \cap D$, a contradiction. Thus we obtain $E^c = E$ and therefore $E \subset c^\perp$. For any line x in E we have $c \in D^x$ and $|D^x| = 2$. Hence $D^x = D$ or $D^x = C$, since every point $X \neq C, D$ on c is connected with A and B by distinct lines $\neq c$ and therefore we have $|X| \geq 3$ for $X \neq C, D$. In every case there is a line $f \neq e$ in E, such that $D^f = D$: If $D^q, D^{q'} \neq D$, then $D^q = C = D^{q'}$ and $D^{eqq'} = D$ and $eqq' \in E$ and $eqq' \neq e$. From $D^f = D$ and $f \notin D$ we deduce $D \subset f^\perp$ and, moreover, $cef = 1$, since $ce \in J$. Since $|E| \geq 3$ (we have $e, q, q' \in E$), there exists a line $g \neq e, f$ in E. Then $D^g \neq D$, and therefore $D^g = C$. This implies $e^g \in C, E$. Thus the Δ-connected point A is not connected with the triangle C, D, E, a contradiction. Thus our assumption that B is 1-Δ-connected is false and (2) is proved. This concludes the proof of Lemma 5.10. ∎

In addition to Lemma 5.10 we show:

PROPOSITION 5.11. *Let* A *be a completely connected point in an S-group plane. Then* $A \cap b^\perp \neq \emptyset$ *for all lines* b *such that* $b \notin A$.

PROOF. By the assumption for A, there exists a line a in $A \cap A^b$. Then $a, a^b \in A$ and $b \notin A$, and therefore $a \in A \cap b^\perp$. ∎

5. THEOREM OF Pappus-Brianchon

The theorems of both Pappus and Desargues can be proved in an S-group plane as long as some of the points in their configurations are completely

---

* Observe that our proof is finished once we know that every point contains at least three lines.

connected. We first prove:

THEOREM 5.12 (Pappus-Brianchon). *Let* a_i, b_k *be lines, and* A, B, C *points of an S-group plane satisfying* $a_i \in A$, $a_i \notin B$, C, *and* $b_k \in B$, *with* B, C *completely connected for* $i, k = 1, 2, 3$. *Moreover, for* $i, k = 1, 2, 3$ *and* $i \neq k$, *let* c_{ik} *be lines such that*

$$c_{ik} \in C \quad \text{and} \quad a_i b_k c_{ik} \in S$$

Then any two of the equations $c_{12} = c_{21}$, $c_{13} = c_{31}$, *and* $c_{23} = c_{32}$ *imply the third.*

PROOF.* It suffices to prove that the equations $c_{12} = c_{21}$ and $c_{13} = c_{31}$ imply $c_{23} = c_{32}$. Since C is completely connected, we can find lines $a, b \in C$ such that $a \in A$ and $b \in B$. Put $\gamma := \gamma_{ab}$ (compare Chapter 4, Section 5)

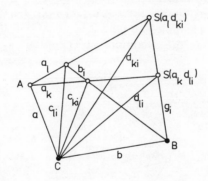

and let $d_{ik} := c_{ik}\gamma$ for $i, k = 1, 2, 3$ and $i \neq k$. Let i, k, l be a cyclic permutation of 1, 2, 3, and let g_i be a line joining B to the point $S(a_k d_{li})$. Note that g_i exists, as B is completely connected. Also, $a_k \neq d_{li}$, as otherwise, $a_k \in C$, contradicting the choice of the $a_i \in A$, $a_i \notin C$. Now observe that the lines a, c_{li}, c_{ki} lie in the domain of $\gamma = \gamma_{ab}$. We can then check that for the lines $b = a\gamma$, $d_{li} = c_{li}\gamma$, $d_{ki} = c_{ki}\gamma$ and the lines b_i, a_k, a_l, and g_i all the assumptions of Theorem 4.2 are fulfilled: This is because $b_i c_{ki} a_k$, $a_k a a_l$ and $a_l c_{li} b_i \in S$, and $b_i b \neq a_k d_{li}$ [since $b_i b = a_k d_{li}$ implies either $b_i = b$ and $a_k = d_{li} \in C$, a contradiction, or $b_i \neq b$ and $a_k \in S(b_i b) = B$, a contradiction]. Finally, we have $b_i b g_i$, $a_k d_{li} g_i \in S$. Thus by Theorem 4.2 we may conclude that $a_l d_{ki} g_i \in S$, hence

$$g_i \in S(a_l d_{ki})$$

* Compare Bachmann [2], p. 73.

We deduce in particular

$$g_1 \in S(a_2 d_{31}), \ S(a_3 d_{21})$$
$$g_2 \in S(a_1 d_{32}), \ S(a_3 d_{12})$$
$$g_3 \in S(a_1 d_{23}), \ S(a_2 d_{13})$$

Now let $l = 1, 2, 3$. Then $g_l \in B$ and $B \neq S(a_k d_{li})$ (since $a_k \notin B$). From $c_{12} = c_{21}$ and $c_{13} = c_{31}$, we obtain $d_{12} = d_{21}$ and $d_{13} = d_{31}$. Thus $g_1, g_2 \in B$, $S(a_3 d_{12})$, and $B \neq S(a_3 d_{12})$. Similarly, $g_1, g_3 \in B$, $S(a_2 d_{13})$, and $B \neq S(a_2 d_{13})$, hence $g_1 = g_2 = g_3$. Moreover, we obtain $g_2, a_1 \in S(a_1 d_{32})$, $S(a_1 d_{23})$, and $a_1 \neq g_2$ (since $a_1 \notin B$ and $g_2 \in B$) and thus $S(a_1 d_{32}) = S(a_1 d_{23})$. Finally, $d_{23}, d_{32} \in C$, $S(a_1 d_{23})$, and $C \neq S(a_1 d_{23})$ (since $a_1 \notin C$), and thus $d_{23} = d_{32}$. But then $c_{23} = d_{23} \gamma^{-1} = d_{32} \gamma^{-1} = c_{32}$, as required. ∎

Note. In the statement of Theorem 5.12 we name three points and a total of 12 lines. For easier reference we introduce the ordered 15-tuple $P(A, B, C; \ a_1, a_2, a_3; \ b_1, b_2, b_3; \ c_{12}, c_{21}, c_{13}, c_{31}, c_{23}, c_{32})$. Whenever we write down such a 15–tuple, we take it to mean that the points and lines so enumerated satisfy the conditions of Theorem 5.12.

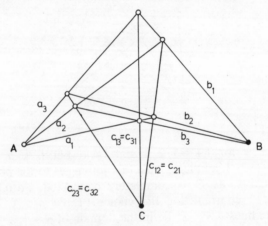

● = completely connected points

6. THEOREM OF Desargues

We now prove the theorem of Desargues, again assuming that some of the points in the configuration are completely connected.

THEOREM 5.13 (Desargues). *Suppose we have the following configuration of points and lines in the S-group plane* $E(G, S)$: g_i, a_i *for* $i = 1$,

2, 3 and o, b_1, b_2 are lines; O, P_i, Q_i, A_i for i = 1, 2, 3 are points; and the following incidences hold: $g_i \in O, P_i, Q_i$; $g_k \notin P_i, Q_i$; $o \in A_i$; $P_i \neq Q_k$ for i, k = 1, 2, 3 and i ≠ k. Also $a_i \in A_i, P_k, P_l$ for any cyclic permutation i, k, l of 1, 2, 3. Also, $b_1 \in A_1, Q_2, Q_3$; $b_2 \in A_2, Q_1, Q_3$; $o \notin O$.

The points P_1, P_2, P_3 and the points Q_1, Q_2, Q_3 are not collinear. Assume that the points O, A_1, A_3 are completely connected. Then there exists a line b_3 such that $b_3 \in A_3, Q_1, Q_2$.

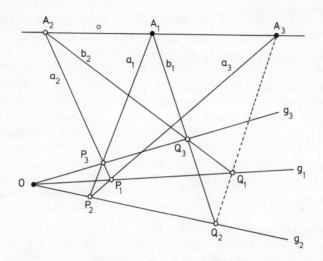

PROOF. To prove this theorem we assume that every line through O (the center of the configuration) is contained in at least six completely connected points. As we see later, this assumption is false only in certain cases where the group G is finite (compare Chapter 7). The finite S-groups with completely connected points are discussed fully in Chapter 8, and we obtain a full proof of Theorem 5.13 from the results there.

In the theory of affine (respectively projective) planes, Hessenberg proved that the theorem of Pappus implies the theorem of Desargues. The proof that follows is essentially a translation of Hessenberg's proof. We omit the minute and elaborate details that are actually required to show that all the appropriate conditions of incidence and nonincidence are fulfilled every time we use the theorem of Pappus-Brianchon, for these discussions follow in exactly the same way as in the projective case. We refer here to reference 11, pp. 204–208.

We observe that in our formulation of the theorem of Desargues we have excluded the case of the little theorem of Desargues.

We now proceed with the proof of Theorem 5.13. Without using the assumption that every line through O is contained in at least six completely connected points, we show the following

(1) *The statement of Theorem 5.13 is true if we replace the condition "O, A_1, A_3 are completely connected" by the condition "P_1, Q_3, A_1, A_3 are completely connected".*

Since A_3 is completely connected, we know that there exist lines b_3', $b_3'' \in A_3$ such that $b_3' \in Q_2$ and $b_3'' \in Q_1$. To prove our assertion, we must show that $b_3' = b_3''$.

We first assume that $g_3 \notin A_3$. We construct points R, M, S as follows. Since $g_3 \in O$ and $o \notin O$, we know $g_3 \neq o$. So we let $R := S(og_3)$. We let r be the line joining the completely connected point P_1 with R. Then $r \neq g_2$, and we let $M := S(rg_2)$. Let m be the line joining the completely connected point A_3 to O. Clearly, $m \neq a_2$, and we put $S := S(ma_2)$. Also, A_1 is completely connected, hence there is a line s joining A_1 to S.

It is now easy to check that we may apply Theorem 5.12 to $P(O, P_1, A_1; g_3, m, g_2; a_3, r, a_2; o, o, a_1, a_1, s, c_{32})$ and conclude that $s \in M$.

Since Q_3 is completely connected, there exists a line $t \in Q_3$, S, and $r \neq t$. We put $T := S(rt)$. Now we apply Theorem 5.12 to the configuration $P(P_1, Q_3, A_3; a_2, g_1, r; g_3, t, b_2; m, m, o, o, b_3'', c_{32})$ and conclude that $b_3'' \in T$.

To apply Theorem 5.12 a third time, we assume $M \neq Q_2$. Then we apply the theorem to the configuration $P(M, Q_3, A_3; s, g_2, r; g_3, t, b_1; m, m, o, o, b_3', b_3'')$ to obtain $b_3' = b_3''$ as required.

If $M = Q_2$, then we obtain $s = b_1$ and $T = M$. But then b_3', $b_3'' \in A_3$, $T = Q_2$ and $A_3 \neq Q_2$; we deduce at once $b_3' = b_3''$.

Next, we suppose $g_3 \in A_3$, but $g_1 \notin A_1$. We then replace the triangles P_1, P_2, P_3 and Q_1, Q_2, Q_3 as follows: we put $P_3' := S(g_1 b_3')$. Let $a_2' := (Q_3, P_3')$ and $A_2' := S(oa_2')$. Let $b_2' := (P_1, A_2')$ and $P_1' := Q_3$, $P_2' := Q_2$. Similarly, let $Q_1' := S(g_3 b_2')$; $Q_2' := P_2$; $Q_3' := P_1$; $g_1' := g_3$; $g_2' := g_2$; $g_3' := g_1$; $A_1' := A_3$; $A_3' := A_1$; $a_1' := b_3'$; $a_3' := b_1$; and $b_1' := a_3$. Then $g_3' \notin A_3'$, and from our discussion above we may deduce that the points Q_1', Q_2', A_3' are connected by a common line. Hence the points P_2, A_1, Q_1' are collinear, from which we deduce successively: $Q_1' = P_3$; $b_2' = a_2$; $A_2' = A_2$; $a_2' = b_2$; $P_3' = Q_1$. Hence the points Q_1, Q_2, A_3 are collinear, which again proves that $b_3' = b_3''$.

Finally, suppose $g_1 \in A_1$ and $g_3 \in A_3$. We assume $b_3' \neq b_3''$ and put $\bar{Q}_1 := S(b_2 b_3')$. Since P_1 is completely connected, there exists a line $\bar{g}_1 \in P_1$, $\bar{Q}_1$. Let $O' := S(\bar{g}_1 g_2)$ and let $\bar{g}_3 \in Q_3$, O'. Next we introduce the following lines and points: $\bar{P}_3 := S(\bar{g}_3 a_2)$; $g_1' := \bar{g}_3$, $g_2' := g_2$; $g_3' := \bar{g}_1$; $a_1' := b_3'$; $a_2' := b_2$; $a_3' := b_1$; $b_1' := a_3$; $b_2' := a_2$; $P_1' := Q_3$; $P_2' := Q_2$; $P_3' := \bar{Q}_1$; $Q_1' := \bar{P}_3$; $Q_2' := P_2$; $Q_3' := P_1$; $A_1' := A_3$; $A_2' := A_2$; and

$A_3' := A_1$. Since $b_3' \neq b_3''$ by assumption, we have $\bar{Q}_1 \neq Q_1$, hence $g_3' \notin A_3'$. Thus we may apply the result proved above to the configuration consisting of the elements g_i', a_i', P_i', Q_i', A_i' ($i = 1, 2, 3$), and b_1', b_2' to deduce that A_3', Q_1', and Q_2' are collinear provided that $o \notin O'$. Thus A_1, P_2, and $\bar{P}_3$ are collinear, which yields in succession: $\bar{P}_3 = P_3$; $\bar{g}_3 = g_3$; $O' = O$; $\bar{g}_1 = g_1$; $\bar{Q}_1 = Q_1$; $b_3'' = b_3'$, in contradiction to our assumption. If $o \in O'$, then we change our construction as follows: Let $Q_1'' := S(b_3'g_1)$ and b_2' the line connecting Q_3 and Q_1'' (this line does exist, since Q_3 is completely connected). Furthermore, we put $A_2' := S(ob_2')$ and denote by a_2' the joining line of A_2' and the completely connected point P_1. By our assumption, $A_2 \neq A_2'$, and therefore $S(g_3a_2') \neq S(a_1a_2')$. Interchanging the roles of the triangles in the old configuration, we see from the same arguments as above that we arrive at a contradiction, or that the intersection O'' of g_2 and the line connecting the points Q_3 and $S(a_1a_2')$ is a point on o, hence $O'' = S(og_2) = O'$. However, in this case we would obtain a contradiction.

This completes the proof of (1).

Next, we show

(2) *The statement of Theorem 5.13 remains true if we replace the assumption "O, A_1, A_3 are completely connected" by the assumption "O, A_1, A_3, Q_3 are completely connected, and* $b_1 \notin P_1$".

Since A_3 is completely connected, we know there exists a line $b_3' \in A_3$, Q_2. Thus $b_3' \neq g_1$, and we may put $\bar{Q}_1 := S(b_3'g_1)$. Similarly, we define $O' := P_1$; $g_1' := g_1$; $g_2' := a_2$; $g_3' := a_3$; $P_1' := O$; $P_2' := P_3$; $P_3' := P_2$; $Q_1' := \bar{Q}_1$; $Q_2' := A_2$; $Q_3' := A_3$; $A_1' := A_1$; $A_2' := Q_2$; $A_3' := Q_3$; $a_1' := a_1$; $a_2' := g_2$; $a_3' := g_3$; $b_1' := o$; $b_2' := b_3'$; $o' := b_1$. Since $b_1 \notin P_1$, we have $o' \notin O'$. Hence we may use (1) for the '-configuration to deduce that Q_1', Q_2', A_3' are collinear. Hence $b_2 \in \bar{Q}_1$, and $\bar{Q}_1 = Q_1$, which implies that Q_1, Q_2, A_3 are collinear, as required.

This completes the proof of (2).

For the remainder of this proof, we assume that every line through O is contained in at least six completely connected points.

Suppose the conditions of Theorem 5.13 are fulfilled. Let b_3' be the line joining the completely connected point A_3 to the point Q_2, and put $Q_1'' := S(b_2b_3')$. We wish to show that $Q_1 = Q_1''$. Choose a completely connected point Q_3' on g_3 satisfying $Q_3' \neq O$, P_3, $o \notin Q_3'$, and the triples P_1, A_1, Q_3', and Q_1'', Q_3', A_1 are both not collinear. This is possible since g_3 contains at least six completely connected points.

Now we can find lines b_1', b_2' such that $b_1' \in Q_3'$, A_1 and $b_2' \in Q_3'$, A_2. We put $Q_2' := S(b_1'g_2)$ and $Q_1' := S(b_2'g_1)$. Since $b_1' \notin P_1$, and since Q_3' is completely connected, we may apply (2) to the triangles P_1, P_2, P_3 and

Q_1', Q_2', Q_3' to deduce the existence of a line $b_3'' \in Q_1'$, Q_2', A_3. Now we have $Q_1'' := S(b_2 b_3')$, and we construct the line g_1' with $g_1' \in O$, Q_1''. Finally, put $Q_1''' := S(g_1' b_2')$. Then $b_1' \notin Q_1''$, and Q_3' is completely connected. We apply (2) to the triangles Q_1'', Q_2, Q_3 and Q_1''', Q_2', Q_3' to deduce the collinearity of Q_1''', Q_2', and A_3. Hence $Q_1''' = Q_1'$ and $g_1' = g_1$, which implies $Q_1'' = Q_1$. Thus Q_1, Q_2, A_3 are collinear, and we are done. ∎

7. EMBEDDABLE S-GROUP PLANES

As we see in Chapter 6, there are very important examples of S-group planes which are projective incidence structures, namely the elliptic planes. However, in general, an S-group plane is not a projective incidence structure—and there exist outstanding examples of such S-group planes: for instance, Euclidean planes, Strubecker planes, hyperbolic-metric planes, and Minkowskian planes (compare Chapter 6). All these S-group planes and many other S-group planes are embeddable in a projective incidence structure. Thus we wish to give a precise definition of the concept of an embeddable S-group plane.

An S-group plane $E(G, S)$ is said to be *embeddable in a projective incidence structure* (L, κ) if up to isomorphisms $E(G, S)$ satisfies the following:

(a) E(G, S) *is a substructure of* (L, κ).

(b) *Any reflection of* E(G, S) *in the line* a *is the restriction of a projective reflection of* (L, κ) *in the line* a.

(A projective reflection σ is defined as an axial collineation such that $\sigma^2 = 1$.) More precisely, we may formulate our conditions for an embeddable S-group plane as follows: The S-group plane $E(G, S)$ is embeddable in the projective incidence structure (L, κ) if there exists an isomorphism Φ of $E(G, S)$ onto a substructure of (L, κ) such that, given any reflection a^* of $E(G, S)$, there exists a collineation α of (L, κ) with axis $a\Phi$ such that

$$x\Phi^{-1} a^* \Phi = x\alpha \qquad \forall x \in S\Phi$$

Clearly, we have $\alpha^2 = 1$. We say $E(G, S)$ is properly embeddable in (L, κ) if the isomorphism Φ of $E(G, S)$ onto a substructure of (L, κ) maps the set of points of $E(G, S)$ onto the set of points of (L, κ).

8. QUADRATIC FORM OF S-GROUP PLANES WITH COMPLETELY CONNECTED POINTS

Let $E(G, S)$ be an S-group plane containing a quadrilateral and at least one completely connected point $S(ab)$ such that $ab \neq ba$. Briefly, we call such a plane an *S-group plane with completely connected points* and any completely connected point $S(ab)$ with $ab \neq ba$ a *proper point*. As we prove in Chapter 7, any S-group plane with completely connected points is properly embeddable in the projective coordinate incidence structure $I(V)$ over a field K. We call K the *coordinate field of the S-group plane* $E(G, S)$. Thus we can regard $E(G, S)$ as a substructure of $I(V)$ and can show:

THEOREM 5.14. *Let* $E(G, S)$ *be an S-group plane with completely connected points regarded as a substructure of the projective coordinate incidence structure* $I(V)$ *over its coordinate field* K. *Then there exists a quadratic form* Q *on* V *such that*

(a) $E(G, S)$ *is a substructure of* $I(V, Q)$.

(b) *Any reflection of* $E(G, S)$ *in the line* a *is a restriction of the reflection of* $I(V, Q)$ *in the line* a.

PROOF. Since we consider $E(G, S)$ to be a substructure of $I(V)$, S is a set of one-dimensional subspaces of V. We put

$$V(S) := \{X \mid X \in V \quad \text{and} \quad \langle X \rangle \in S\}$$

Let $A \in V(S)$ and $a := \langle A \rangle$. By our assumptions, there is an axial collineation α_a of $(L(V), \kappa(V))$ with axis a such that

$$\langle X \rangle \alpha_a = \langle X \rangle a^* \qquad \forall X \in V(S)$$

We have $\alpha_a^2 = 1$. Since any such collineation is induced by a linear map, there is a linear map σ in $GL(V)$ such that

$$\langle X\sigma \rangle = \langle X \rangle \alpha_a \qquad \forall X \neq O \quad \text{in} \quad V$$

Since a^* and thus α_a fixes the line a, and since the linear map inducing α_a is only determined up to a factor, we may assume $A\sigma = A$. In this case we write σ_A instead of σ.

From $\alpha_a^2 = 1$ we obtain $X\sigma_A^2 = rX$ for all X in V with a suitable r in K. From $A = A\sigma_A^2 = rA$ we deduce $r = 1$, hence we have $\sigma_A^2 = 1$.

Next we prove the following

(1) *There is a map* $\lambda : V \times V(S) \to K$ *such that*

$$X\sigma_A = -X + \lambda(X, A)A \qquad \forall X \text{ in } V.$$

Suppose there is a vector X in V such that $X\sigma_A + X \notin \langle A \rangle$. Then X, A are linearly independent and therefore $\dim \langle X, A \rangle = 2$. Since $\langle A \rangle$ is an axis of α_a, we have $X\sigma_A \in \langle X, A \rangle$ and therefore $\langle X, A \rangle = \langle X\sigma_A + X, A \rangle$. The linear map σ_A fixes A and $X\sigma_A + X$, since $(X\sigma_A + X)\sigma_A = X + X\sigma_A$. Thus every vector in $\langle X, A \rangle$ is fixed. Especially, $X\sigma_A = X$ and therefore

$X\sigma_A + X = 2X$. From $X\sigma_A + X \notin \langle A \rangle$ we see $2X \neq O$ and thus char $K \neq 2$. Thus the axis of the axial collineation α_a of a Pappian projective incidence structure of char $\neq 2$ is incident with a center of α_a, namely with the center $\langle X, A \rangle$, and therefore we obtain $\alpha_a = 1$ (compare the Appendix). This implies $a^* = 1$ and $a \in Z(G)$. For any $b \neq a$ in S we then have $b^* \neq 1$ (because of Corollary 5.3) and therefore $\alpha_b \neq 1$. But since $S(ab)$ is a center of b^* and the axis b of b^* is incident with this center, we arrive at the contradiction $\alpha_b = 1$.

Thus we have proved $X\sigma_A + X \in \langle A \rangle$ for all X in V. This yields (1).

Our theorem is proved if we can show that there is a quadratic form Q on V such that $Q(A) \neq 0$ and $\lambda(X, A) = (g_Q(X, A))/Q(A)$ for all X in V and all A in $V(S)$.

To elaborate this, we deduce some immediate consequences of the representation (1) for the reflection a^* in the line a.

From $A\sigma_A = A$ we deduce

(2) $\lambda(A, A) = 2 \quad \forall A \in V(S).$

(3) σ_A *fixes the line* $\langle X \rangle \Leftrightarrow \langle X \rangle = \langle A \rangle$ *or* $\lambda(X, A) = 0$.

(4)   *If* $A, B \in V(S)$ *and* $\langle A \rangle \neq \langle B \rangle$, *then*
$\lambda(A, B) = 0 \Leftrightarrow \lambda(B, A) = 0$.

We put $a := \langle A \rangle$ and $b := \langle B \rangle$ and deduce from (3): $\lambda(A, B) = 0 \Leftrightarrow b^*$ fixes $a \Leftrightarrow ab = ba \Leftrightarrow a^*$ fixes $b \Leftrightarrow \lambda(B, A) = 0$.

(5)   *For* A *in* $V(S)$ *the map*

$$\lambda_A : \begin{cases} V \to K \\ X \mapsto \lambda(X, A) \end{cases}$$

is linear.

(6) $\sigma_A = \sigma_{xA} \; \forall x \neq 0$ *in* K.

Both maps σ_A and σ_{xA} induce the collineation α_a with $a := \langle A \rangle$. From $A\sigma_A = A = (x^{-1})(xA) = x^{-1}(xA)\sigma_{xA} = A\sigma_{xA}$ we see that $\sigma_A = \sigma_{xA}$.

From (1) and (6) we deduce immediately:

(7) $\lambda(X, xA) = x^{-1}\lambda(X, A) \quad \forall (X, A) \in V \times V(S); \quad \forall x \neq 0$ *in* K.

(8) $\sigma_A \sigma_B \sigma_A = \sigma_{B\sigma_A} \quad \forall A, B$ *in* $V(S)$.

This follows for $a := \langle A \rangle$ and $b := \langle B \rangle$ from $a^* b^* a^* = (ba^*)^*$.

From (1), (7), and (8) one can easily compute the following two statements:

(9) $\lambda(X\sigma_A, B\sigma_A) = \lambda(X, B) \quad \forall X \in V$ *and* $A, B \in V(S)$.

(10) $X\sigma_A \sigma_B = X - \lambda(X, A)A - \lambda(X, B)B + \lambda(X, A)\lambda(A, B)B \; \forall X \in V$ *and* $A, B \in V(S)$.

(11)   *Let* $a, b, c, d \in S$ *and* $\langle A \rangle := a; \langle B \rangle := b; \langle C \rangle := c; and \langle D \rangle := d$. *Then* $abc = d$ *implies* $\sigma_A \sigma_B \sigma_C = \sigma_D$.

From $abc = d$ we deduce $\alpha_a \alpha_b \alpha_c = \alpha_d$ for the collineations α_a, α_b, α_c, α_d induced by a, b, c, d. Thus there is an element s in K such that $\sigma_A \sigma_B \sigma_C = s\sigma_D$. The lines a, b, c, d are concurrent, hence there is a two-dimensional subspace T of V containing A, B, C, D. From (1) we obtain $X\sigma_A\sigma_B\sigma_C = -X + Y$ and $Y \in T$ for all vectors X in V (obviously, the vector Y depends on the vector X). Hence we have $-X + Y = s(-X + \lambda(X, D)D)$ and therefore $X(1-s) \in T$ for all X in V. This yields $s = 1$, and (11) is proved.

Next, we consider the equation

$$\lambda(A, B)\lambda(B, C)\lambda(C, A) = \lambda(B, A)\lambda(C, B)\lambda(A, C) \qquad (*)$$

for suitable vectors A, B, C in $V(S)$, and prove

(12) $(*)$ *holds, if* A, B, C *are linearly dependent vectors in* V(S).

Obviously, $(*)$ follows from (10) and the following equation:

$$\lambda(C\sigma_A\sigma_B - C\sigma_B\sigma_A, C) = 0 \qquad (**)$$

Thus, if we wish to prove $(*)$ for a triplet A, B, C of vectors in $V(S)$, we only need to prove $(**)$. Now let A, B, C be any linearly dependent vectors in $V(S)$. Then for $a := \langle A \rangle$, $b := \langle B \rangle$, $c := \langle C \rangle$ we have $abc \in S$. From (11) and $\sigma_X^2 = 1$ for all X in $V(S)$ we deduce $\sigma_A\sigma_B\sigma_C = \sigma_C\sigma_B\sigma_A$ and obtain

$$\lambda(C\sigma_A\sigma_B - C\sigma_B\sigma_A, C) = \lambda(C\sigma_A\sigma_B - C\sigma_C\sigma_B\sigma_A, C)$$
$$= \lambda(C\sigma_A\sigma_B - C\sigma_A\sigma_B\sigma_C, C)$$
$$\overset{(5)}{=} \lambda(C\sigma_A\sigma_B, C) - \lambda(C\sigma_A\sigma_B\sigma_C, C)$$
$$= \lambda(C\sigma_A\sigma_B, C) - \lambda(C\sigma_A\sigma_B\sigma_C, C\sigma_C)$$
$$\overset{(9)}{=} \lambda(C\sigma_A\sigma_B, C) - \lambda(C\sigma_A\sigma_B, C) = 0.$$

Thus $(**)$, and therefore $(*)$, is valid for A, B, C.

(13) $(*)$ *holds if* A, B, C *are linearly independent vectors in* V(S) *and if there exists at least one altitude in the triangle* $\langle A \rangle$, $\langle B \rangle$, $\langle C \rangle$.

Without loss of generality we may assume that there exists an altitude $\langle D \rangle$ in the triangle $\langle A \rangle$, $\langle B \rangle$, $\langle C \rangle$ to the line $\langle C \rangle$. By (11) there is a line $\langle E \rangle$ such that $\sigma_D\sigma_A\sigma_B = \sigma_E$. Since $\langle C \rangle$ is orthogonal to $\langle D \rangle$, we obtain from (4) and (1) $\lambda(C, D) = 0$ and $C\sigma_D = -C$ and therefore $C\sigma_A\sigma_B - C\sigma_B\sigma_A = C\sigma_D\sigma_E - C\sigma_E\sigma_D = -C\sigma_E - C\sigma_E\sigma_D = -\lambda(C\sigma_E, D)D$. This implies $\lambda(C\sigma_A\sigma_B - C\sigma_B\sigma_A, C) \overset{(5)}{=} -\lambda(C\sigma_E, D)\lambda(D, C) = 0$. Thus $(**)$, and therefore $(*)$, is valid for the vectors A, B, C.

If at least one edge of a triangle is a completely connected point A, then there is at least one altitude in that triangle, since there is a line through A which is orthogonal to the opposite side of the triangle (compare Proposition 5.11). Thus we can apply (13) for any triangle in S

for which at least one edge is completely connected. We don't refer to this fact explicitly in the following.

We introduce the following functions:

$$q_A(B) := \begin{cases} \lambda(B, A)\lambda(A, B)^{-1} & \text{if } A, B \in V(S) \text{ and } \lambda(A, B) \neq 0 \\ x^2 & \text{if } B = xA \in V(S) \end{cases}$$

$$G_A(B, C) := q_A(B)\lambda(C, B) \quad \text{if } A, B \in V(S); \ C \in V; \text{ and}$$
$$\text{if } q_A(B) \text{ is defined}$$

Our definition of $q_A(B)$ is correct, since $B = xA \in V(S)$ and $\lambda(A, B) \neq 0$ imply by (5) and (7) $\lambda(B, A)\lambda(A, B)^{-1} = x^2$. Obviously, we obtain from (5) and (7):

(14) *For the functions* $q_A(B)$ *and* $G_A(B, C)$ *we have* $q_A(xB) = x^2 q_A(B)$; $G_A(xB, C) = xG_A(B, C) \ \forall x \neq 0 \ in \ K. \ q_{xA}(B) = x^{-2} q_A(B) \ \forall x \neq 0 \ in \ K.$ *The map:* $X \mapsto G_A(B, X)$ *of* V *into* K *is linear.*

Moreover, our definitions immediately imply:

(15) *If* A, B, C *in* V(S) *satisfy the equation* (∗) *and if* $q_A(B)$, $q_B(C), q_A(C)$ *are defined, then* $q_A(B)q_B(C) = q_A(C)$.

Next we prove

(16) *If* A, B, C *in* V(S) *satisfy the equation* (∗) *and if* $q_A(B)$ *and* $q_A(C)$ *are defined, then* $G_A(B, C) = G_A(C, B)$.

We compute for $\lambda(A, B) = 0$ and therefore $B = xA$ that

$$G_A(B, C) = x^2 \lambda(C, B) = x\lambda(C, A)$$
$$= \begin{cases} xq_A(C)\lambda(A, C) & \text{if } \lambda(A, C) \neq 0 \\ xy^2\lambda(A, C) & \text{if } C = yA \end{cases} = q_A(C)\lambda(xA, C)$$
$$= q_A(C)\lambda(B, C) = G_A(C, B).$$

If $\lambda(A, B) \neq 0$, then we have

$$G_A(B, C) = \lambda(B, A)\lambda(A, B)^{-1}\lambda(C, B)$$
$$= \begin{cases} \lambda(C, A)\lambda(A, C)^{-1}\lambda(B, C) & \text{if } \lambda(A, C) \neq 0 \\ y^2\lambda(B, C) & \text{if } C = yA \end{cases}$$
$$= q_A(C)\lambda(B, C) = G_A(C, B).$$

Thus (16) is proved.

(17) *Let* A, B *be two vectors in* V(S) *such that* $q_A(B)$ *and* $q_A(A+B)$ *are defined. Then*

$$q_A(A + B) = q_A(A) + q_A(B) + G_A(A, B)$$

If A, B are linearly dependent, then there is an element $x \neq 0$ in K such that $B = xA$, and our assertion follows immediately from the definitions and from (2) and (14). Now let A, B be linearly independent vectors in $V(S)$, and put $a := \langle A \rangle$, $b := \langle B \rangle$, and $c := \langle A + B \rangle$. By our assumptions $q_A(A+B)$ is defined, thus $A + B \in V(S)$, and a, b, c are concurrent lines in S. Hence there is a line d in S such that $abc = d$. We put $\langle D \rangle := d$ and obtain from (11) $\sigma_B \sigma_A \sigma_{A+B} = \sigma_D$. The vector D is a linear combination of A and B, and we have $\langle B \rangle \neq \langle D \rangle$ (otherwise $\langle A + B \rangle = \langle A \rangle$ and A, B would be linearly dependent). Thus there is an element u in K such that $D = A + uB$. Applying (10) to $\sigma_B \sigma_A = \sigma_D \sigma_{A+B}$ and using the fact that A, B are linearly independent, we obtain

$$u\lambda(B, A) + \lambda(B, A)\lambda(A, B) + \lambda(A, B) = 0 \tag{1}$$

Analogously, we deduce from $\sigma_{B+A}\sigma_{-A} = \sigma_{\bar{D}}\sigma_B$ and $\bar{D} = [u/(u-1)](B+A) + (-A)$ for $\bar{D} = [1/(u-1)]D$ (we have $u - 1 \neq 0$, otherwise $\langle D \rangle = \langle A + B \rangle$ and $\langle A \rangle = \langle B \rangle$ in contradiction to the fact that A, B are linearly independent) and $B = (B + A) + (-A)$ the following equation:

$$\frac{u}{u-1}\lambda(B+A, -A) + \lambda(B+A, -A)\lambda(-A, B+A) + \lambda(-A, B+A) = 0$$

$$\tag{2}$$

By our assumptions $q_A(B)$ is defined and therefore $\lambda(A, B) \neq 0$, since $\langle A \rangle \neq \langle B \rangle$. Combining equations 1 and 2, we obtain

$$\lambda(A, A+B) = \frac{\lambda(A+B, A)\lambda(A, B)}{\lambda(B, A) + \lambda(B, A)\lambda(A, B) + \lambda(A, B)}$$

This equation yields (17) if we apply the definition of $q_A(A+B)$ and $q_A(A)$ and $q_A(B)$.

To introduce a suitable basis for the vector space V we choose a triangle e_1, e_2, e_3 in S such that $e_1 \pm e_k$ for i, $k = 1, 2, 3$ and such that for every vertex of e_1, e_2, e_3 there is an altitude in the triangle. Moreover, we assume that $S(e_i e_k)$ for i, $k = 1, 2, 3$ and $i \neq k$ is a proper point. That choice is possible if $E(G, S)$ is infinite, as we see from Chapter 7, Section 2. If $E(G, S)$ is finite, then we deduce from Chapter 8, Section 2, that we can find such a triangle with the only exception that $|A| = 4$. However, then $E(G, S)$ is a hyperbolic-metric plane or a Euclidean plane, and the assertion of Theorem 5.13 follows by a direct argument (compare also Chapter 6, Section 7, which is independent of Theorem 5.13).

Now, let $\langle E_i \rangle := e_i$ for $i = 1, 2, 3$. Then E_1, E_2, E_3 is a basis of V. Moreover, $\lambda(E_i, E_k) \neq 0$ if $i \neq k$ and therefore the elements $q_{E_i}(E_k)$ are defined for any pair i, k with $1 \leq i, k \leq 3$. We put $q(X) := q_{E_1}(X)$ and $G(X, Y) := G_{E_1}(X, Y)$ if $q_{E_1}(X)$ is defined. Obviously, $G(E_i, E_k)$ are defined for $1 \leq i, k \leq 3$.

We consider the quadratic form

$$Q : \begin{cases} V \to K \\ \displaystyle\sum_{i=1}^{3} x_i E_i \mapsto \sum_{i=1}^{3} x_i^2 q(E_i) + \sum_{i<k} x_i x_k G(E_i, E_k) \end{cases}$$

and put $g := g_Q$. We prove the following

(18)   *For all* i, k *we have* $G(E_i, E_k) = g(E_i, E_k)$.

If $i = k$, we obtain from (2) $G(E_i, E_i) = q(E_i)\lambda(E_i, E_i) = 2q(E_i) = 2Q(E_i) = g(E_i, E_i)$ (compare Chapter 1). If $i < k$ and $1 \le i, k \le 3$, our definition of the quadratic form Q yields $G(E_i, E_k) = Q(E_i + E_k) - Q(E_i) - Q(E_k) = g(E_i, E_k)$. Finally, if $i > k$ and $1 \le i, k \le 3$, then we obtain, applying (16), $G(E_i, E_k) = G(E_k, E_i) = g(E_k, E_i) = g(E_i, E_k)$, since there exists at least one altitude in the triangle $\langle E_1 \rangle, \langle E_2 \rangle, \langle E_3 \rangle$. Thus, (18) is proved.

(19)   *We have* $G(E_i, X) = g(E_i, X)$ *for all* X *in* V *and* $i = 1, 2, 3$.

We deduce the following from (18) and (14):

$$\begin{aligned} G(E_i, X) &= G(E_i, \textstyle\sum_{k=1}^{3} x_k E_k) \\ &= \sum x_k G(E_i, E_k) = \sum x_k g(E_i, E_k) \\ &= g(E_i, \sum x_k E_k) = g(E_i, X) \end{aligned}$$

for all X in V.

(20)   *Let* $X = x_i E_i + x_k E_k$ *be a vector in* $V(S)$ *and* $i \ne k$. *Then* $q_{E_i}(X)$ *or* $q_{E_k}(X)$ *is defined. If* $q_{E_s}(X)$ *is defined for* $1 \le s \le 3$, *then* $Q(X) = q(E_s) q_{E_s}(X)$.

Since any line in $S(e_i e_k)$ is orthogonal to at most one of the lines e_i, e_k [since $S(e_i e_k)$ is a proper point; compare Chapter 7, Section 2(1)], we have $\lambda(E_i, X) \ne 0$ or $\lambda(E_k, X) \ne 0$. Hence $q_{E_i}(X)$ or $q_{E_k}(X)$ is defined.

Now let $q_{E_s}(X)$ be defined for any element s such that $1 \le s \le 3$. If $x_i = 0$, then $x_k \ne 0$ and, by (14), (15), we obtain

$$\begin{aligned} q(E_s) q_{E_s}(X) &= q(E_s)[q_{E_s}(x_k E_k)] \\ &= q(E_s)[x_k^2 q_{E_s}(E_k)] \\ &= x_k^2 q(E_s) q_{E_s}(E_k) \\ &= x_k^2 q(E_k) = Q(X), \end{aligned}$$

as required. Analogously, we can compute $Q(X) = q(E_s) q_{E_s}(X)$ if $x_k = 0$. Thus we can assume $x_i, x_k \ne 0$ and $i < k$. Then we deduce from (12), (13),

(15) and (17) in the case $s = i$ or $s = k$:

$$
\begin{aligned}
Q(X) &= x_i^2 q(E_i) + x_k^2 q(E_k) + x_i x_k G(E_i, E_k) \\
&= q(x_i E_i) + q(x_k E_k) + G(x_i E_i, x_k E_k) \\
&= q(x_s E_s) q_{x_s E_s}(x_i E_i) + q(x_s E_s) q_{x_s E_s}(x_k E_k) \\
&\quad + q(x_s E_s) G_{x_s E_s}(x_i E_i, x_k E_k) \\
&= q(x_s E_s)[q_{x_s E_s}(x_i E_i) + q_{x_s E_s}(x_k E_k) \\
&\quad + G_{x_s E_s}(x_i E_i, x_k E_k)] \\
&= q(x_s E_s) q_{x_s E_s}(x_i E_i + x_k E_k) \\
&= q(x_s E_s) q_{x_s E_s}(X) = q(E_s) q_{E_s}(X),
\end{aligned}
$$

as required.

If $s \neq i, k$, then our proof shows that without loss of generality $Q(X) = q(E_i) q_{E_i}(X)$. Then by (13) and (15) we obtain $Q(X) = q(E_i) q_{E_i}(X) = q(E_i) q_{E_i}(E_s) q_{E_s}(E_i) q_{E_i}(X) = q(E_s) q_{E_s}(X)$, since there is at least one altitude in the triangle $\langle E_s \rangle$, $\langle E_i \rangle$, $\langle X \rangle$. This proves (20).

From now on, we use the assumption that $S(e_i e_k)$ is a proper point for all i, k such that $i \neq k$ and $i, k = 1, 2, 3$.

(21) Let $X = x_1 E_1 + x_2 E_2 + x_3 E_3$ be any vector in $V(S)$ such that $\langle X \rangle$ is not a central line. Then there exists at least one number i such that $1 \leq i \leq 3$ and $q_{E_i}(X)$ is defined. Moreover, if $q_{E_s}(X)$ is defined for an integer s such that $1 \leq s \leq 3$, then $Q(X) = q(E_s) q_{E_s}(X)$.

If we assume that $q_{E_i}(X)$ is not defined for $i = 1, 2, 3$, then $\lambda(E_i, X) = 0$ for $i = 1, 2, 3$, and (5) yields $\lambda(Y, X) = 0$ for all Y in V; therefore $\langle X \rangle$ would be a central line. Thus there is at least one number i such that $1 \leq i \leq 3$ and $q_{E_i}(X)$ is defined.

Now let $q_{E_s}(X)$ be defined. If x_1, x_2, x_3 are not all $\neq 0$, then our assertion follows from (20). Thus we may assume $x_1, x_2, x_3 \neq 0$. Then, especially, $\langle X \rangle \neq \langle E_s \rangle$, hence $\lambda(E_s, X) \neq 0$. Let E_r and E_t be the distinct vectors $\neq E_s$. We put $A := x_r E_r + x_t E_t$, then $A \in V(S)$, since $S(e_r e_t)$ is a completely connected point. First, let $\lambda(A, E_s) \neq 0$; then $q_{E_s}(A)$ is defined, and therefore $q_{x_s E_s}(A)$ is defined, and we have, by (14) $q_{x_s E_s}(A) = (1/x_s)^2 \cdot q_{E_s}(A)$. From (13), (14), (15), (17), (19), and (20) we deduce

$$
\begin{aligned}
q(E_s) q_{E_s}(X) &= q(x_s E_s) q_{x_s E_s}(X) \\
&= q(x_s E_s)[q_{x_s E_s}(x_s E_s) + q_{x_s E_s}(A) + G_{x_s E_s}(x_s E_s, A)] \\
&= q(x_s E_s) + q(E_s) q_{E_s}(A) + G(x_s E_s, A) \\
&= Q(x_s E_s) + Q(A) + g(x_s E_s, A) = Q(x_s E_s + A) = Q(X)
\end{aligned}
$$

as required.

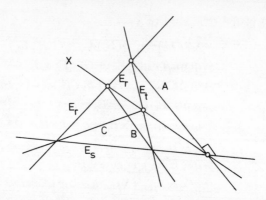

If $\lambda(A, E_s) = 0$, then $\lambda(X, E_r)$ and $\lambda(X, E_t)$ are not both zero [otherwise we would obtain $\lambda(A, X) = 0$ and therefore $\lambda(X, E_s) = 0$, a contradiction]. Without loss of generality we may assume $\lambda(X, E_r) \neq 0$, hence $q_{E_r}(X)$ is defined. We put $B := x_s E_s + x_t E_t$. First, let $\lambda(B, E_r) \neq 0$. Then the above proof shows $Q(X) = q(E_r)q_{E_r}(X)$, and from (13), (15), and (17) we deduce $Q(X) = q(E_r)q_{E_r}(X) = q(E_r)q_{E_r}(E_s)q_{E_s}(E_r)q_{E_r}(X) = q(E_s)q_{E_s}(X)$, as required. If $\lambda(B, E_r) = 0$ and $\lambda(X, E_t) \neq 0$, then $\lambda(C, E_t) \neq 0$ for $C := x_r E_r + x_s E_s$, since we conclude from Theorem 4.3 that the altitudes in the triangle $\langle E_1 \rangle$, $\langle E_2 \rangle$, $\langle E_3 \rangle$ are concurrent, hence $\lambda(C, E_t) = 0$ would imply char $K = 2$ (the complete quadrangle $\langle E_s \rangle$, $\langle E_r \rangle$, $\langle E_t \rangle$, $\langle A \rangle$, $\langle B \rangle$, $\langle C \rangle$ has collinear diagonal points), and therefore the center of the collineation induced by σ_{E_s} would be incident with the axis $\langle E_s \rangle$ and with $\langle A \rangle$, and thus $\lambda(X, E_s) = 0$, a contradiction. Hence the proof just given above implies $Q(X) = q(E_t)q_{E_t}(X) = q(E_s)q_{E_s}(X)$, as before. The last case, $\lambda(B, E_r) = 0$ and $\lambda(X, E_t) = 0$, is possible for at most one line $\langle X \rangle$ in $\langle A, E_s \rangle$. In particular, these conditions are not fulfilled for the line $\langle Y \rangle$ with $Y := X\sigma_{E_s}$, since $\langle Y \rangle \neq \langle X \rangle$. Moreover, we have $\langle Y \rangle \neq \langle A \rangle, \langle E_s \rangle$ and our above arguments show that $q_{E_s}(Y)$ is defined and that $Q(Y) = q(E_s)q_{E_s}(Y)$. There exists an element $u \neq 0$ in the field K such that $Y = X + uE_s$. From (17) we obtain

$$q_{uE_s}(Y) = q_{uE_s}(X) + q_{uE_s}(uE_s) + G_{uE_s}(X, uE_s)$$

and therefore, by (14),

$$q_{E_s}(Y) = q_{E_s}(X) + q_{E_s}(uE_s) + G_{E_s}(X, uE_s)$$

This implies

$$Q(X) + Q(uE_s) + g(X, uE_s) = Q(Y) =$$
$$q(E_s)q_{E_s}(Y) = q(E_s)[q_{E_s}(X) + q_{E_s}(uE_s) + G_{E_s}(X, uE_s)]$$
$$= q(E_s)q_{E_s}(X) + Q(uE_s) + g(X, uE_s)$$

and therefore $Q(X) = q(E_s)q_{E_s}(X)$, as required. Thus (21) is proved.

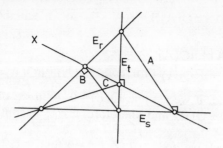

(22) *If* $X \in V(S)$ *and* $\langle X \rangle$ *is not a central line, then* $Q(X) \neq 0$ *and* $\lambda(Y, X) = Q(X)^{-1} g(Y, X)$ *for all* Y *in* V.

By (21) we have $Q(X) \neq 0$. To prove the second statement, it suffices to show $\lambda(E_i, X) = Q(X)^{-1} g(E_i, X)$ for $i = 1, 2, 3$ [compare (5)]. If $\lambda(E_i, X) = 0$ for a number i such that $1 \leq i \leq 3$, then (4) implies $\lambda(X, E_i) = 0$ and therefore, by (19), $g(E_i, X) = G(E_i, X) = q(E_i) \lambda(X, E_i) = 0$. This yields $\lambda(E_i, X) = Q(X)^{-1} g(E_i, X)$, as required. Now, let $\lambda(E_i, X) \neq 0$, then $q_{E_i}(X)$ is defined, and from (21) we see that $Q(X) = q(E_i) q_{E_i}(X)$. Thus we can compute, applying (21), (12), (16), and (19),

$$\lambda(E_i, X) = q_{E_i}(X)^{-1} G_{E_i}(X, E_i) = [q(E_i) q_{E_i}(X)]^{-1} q(E_i) G_{E_i}(X, E_i)$$
$$= Q(X)^{-1} q(E_i) G_{E_i}(E_i, X) = Q(X)^{-1} G(E_i, X) = Q(X)^{-1} g(E_1, X),$$

as required. Thus (22) is proved.

(23) *If* $Z \in V(S)$ *and* $\langle Z \rangle$ *is a central line, then* $Q(Z) \neq 0$ *and* $\lambda(X, Z) = g(X, Z)/Q(Z)[= 0]$ *for all* X *in* V.

We put $z := \langle Z \rangle$ and choose any two distinct lines $a, b \neq z$ such that $abz \in S$. Let $d := abz$ and $\langle A \rangle := a$ and $\langle B \rangle := b$ and $\langle D \rangle := d$. Without loss of generality we may assume $Z = A + B$. From $\lambda(X, Z) = 0$ for all X in $V(S)$ such that $X \notin \langle Z \rangle$ we obtain $\lambda(E_i, Z) = 0$ for $i = 1, 2, 3$ and therefore $2 = \lambda(Z, Z) = 0$. Thus we have char $K = 2$ and $\lambda(X, Z) = 0$ for all X in V and $\sigma_Z = 1$. By (11), we have $\sigma_A \sigma_B = \sigma_A \sigma_B \sigma_Z = \sigma_D$. This equation yields, after (10) is applied,

$$\frac{g(X, A)}{Q(A)} A + \frac{g(X, B)}{Q(B)} B = \frac{g(X, D)}{Q(D)} D \qquad \forall X \in V \qquad (3)$$

since $d \perp z$, and therefore $a \perp b$ and $\lambda(A, B) = 0$. From $D \in \langle A, B \rangle$ and $\langle D \rangle \neq \langle Z \rangle$ we deduce that there is an element u in K such that $D = uA + B$ and $u \neq 1$. Since A, B are linearly independent, equation 3 yields $g(X, A) Q(D) = g(X, D) Q(A) u$, hence $g(X, Q(D) A + uQ(A) D) = 0$ for all X in V, and therefore $Q(D) A + uQ(A) D \in \langle Z \rangle$. This implies

$(Q(D) + u^2 Q(A))A + uQ(A)B = s(A + B)$ for a suitable s in K, and therefore

$$0 = Q(D) + u^2 Q(A) + uQ(A)$$
$$= u^2 Q(A) + Q(B) + u^2 Q(A) + uQ(A) = uQ(A) + Q(B).$$

On the other hand, if we assume $Q(Z) = 0$, then we obtain $0 = Q(Z) = Q(A + B) = Q(A) + Q(B)$ and therefore $u = 1$, a contradiction. Hence $Q(Z) \neq 0$.

From $\lambda(E_i, Z) = 0$ we deduce $\lambda(Z, E_i) = 0$ for $i = 1, 2, 3$ by (4). Hence we have $G(E_i, Z) = q(E_i)\lambda(Z, E_i) = 0$ for $i = 1, 2, 3$ and therefore for all $X = \sum_{i=1}^{3} x_i E_i$ in V; applying (19), we obtain

$$g(X, Z) = \sum_{i=1}^{3} x_i g(E_i, Z) = \sum_{i=1}^{3} x_i G(E_i, Z) = 0$$

and therefore $\lambda(X, Z) = 0 = Q(Z)^{-1} g(X, Z)$ for all X in V. Thus (23) is proved.

From (22) and (23) we see the following

(24) *For any* $a = \langle A \rangle$ *in* S *we have* $Q(A) \neq 0$; *thus* $a \in I(V, Q)$, *and the reflection* a^* *is the restriction of the reflection* $\bar\sigma_A$ *of* $I(V, Q)$ *in the line* a.

This proves Theorem 5.14. ■

6
COMPLETE S-GROUP PLANES

1. MAIN THEOREMS ON COMPLETE S-GROUP PLANES

An S-group plane is called a *complete S-group plane* if it is Δ-connected and if it contains a quadrilateral. Moreover, the group planes of the following two exceptional S-groups (G, S) are also called complete S-group planes:

1. $(G, S) = (G_0, S_0)$ such that S_0 contains exactly two elements.
2. $(G, S) = (G_1, S_1)$ such that S_1 contains exactly four nonconcurrent lines, three of them concurrent and the other orthogonal to the others.

We recall that the assumptions on the existence of lines made above are necessary only in the case that all the points of the S-group plane are completely connected since, if there exists at least one point not completely connected, then there exists a quadrilateral.

Our goal in this chapter is to prove the following theorems characterizing the complete S-group planes and their underlying S-groups.

MAIN THEOREM 6.1. *Every complete S-group plane is isomorphic to the incidence structure over a suitable three-dimensional metric vector space* (V, Q) *such that dim* $V^{\perp} \leq 2$.

Conversely, any incidence structure $I(V, Q)$ *over a three-dimensional metric vector space* (V, Q) *such that dim* $V^{\perp} \leq 2$ *is isomorphic to a suitable complete S-group plane.*

MAIN THEOREM 6.2. *Let* (G, S) *be an S-group distinct from* (G_0, S_0) *such that* $E(G, S)$ *is a complete S-group plane and let* $I(V, Q)$ *be the* *incidence structure over a metric vector space* (V, Q) *over a field* K *which is isomorphic to* $E(G, S)$ *according to Theorem 6.1.*

Then (G^*, S^*) *is isomorphic to the orthogonal group* $(O(V, Q), S(V, Q))$, *and the center* $Z(G)$ *is equal to* $\{1\}$ *or to the cyclic group of order* 2 *or isomorphic to the additive group* K^+ *of the field* K.

The latter theorem illustrates the structure of any S-group $(G, S) \neq (G_0, S_0)$ whose group plane is a complete S-group plane, since G^* is isomorphic to the factor group $G/Z(G)$.

To prove our two main theorems, we may exclude the two exceptional cases, since it is easily verified that $E(G_0, S_0)$ is isomorphic to the incidence structure $I(V, Q)$ such that $V := (GF(2))^3$ and dim rad $(V, Q) = 1$ and Ind $(V, Q) = 1$, and that $E(G_1, S_1)$ is isomorphic to the incidence structure $I(V, Q)$ such that $V := (GF(2))^3$ and dim rad $(V, Q) = 1$ and Ind $(V, Q) = 0$. Moreover, in the latter case we have $(G_1^*, S_1^*) \cong (O(V, Q), S(V, Q))$, and $Z(G)$ is the cyclic group of order two, since G_1^*, as well as $O(V, Q)$, is the dihedral group of six elements, and since S_1^* as well as $S(V, Q)$ is the set of the elements not in the abelian normalizer of index 2.

To prepare a proof of our theorems, we collect some facts on 2-Δ-connected points.

2.   2-Δ-CONNECTED POINTS IN AN ARBITRARY S-GROUP PLANE

The first statements do not use the assumption that the S-group plane is Δ-connected.

LEMMA 6.3. *Let* A *denote any point in an arbitrary S-group plane containing a quadrilateral. Then* A *is* 2-Δ-*connected if and only if for any line* b *not in* A *there exists exactly one point* B *on* b *not connected with* A.

Moreover, if A *is* 2-Δ-*connected and* B *is a point on* b *not connected with* A *and* $a \in A \cap b^{\perp}$, *then* $B \subset a^{\perp}$.

PROOF.   Let A be any point such that for any line b not in A there exists exactly one point B on b not connected with A. Then A is Δ-connected, by definition. Since there exists a quadrilateral, we can find a line b such that $b \notin A$. This implies that A is not completely connected, and therefore 2-Δ-connected, by definition.

Conversely, let A be any 2-Δ-connected point and let b be any line not in A. Then Corollary 5.9 yields that there exists at least one point B on b

not connected with A. Since A is 2-Δ-connected, there is no other point on b not connected with A.

If $a \in A \cap b^{\perp}$, then B, B^a are two points on b not connected with A. Hence $B^a = B$ and $a \notin B$ (since A, B are not connected) and therefore $B \subset a^{\perp}$. ∎

LEMMA 6.4. *If an S-group plane contains a central line, then that line is contained in every 2-Δ-connected point.*

PROOF. Let z be a central line and A be a 2-Δ-connected point. We assume that $z \notin A$ and choose two distinct lines a, a' in A. From Lemma 6.3 we deduce that there is exactly one point B on z not connected with A and that $B \subset a^{\perp}$ and $B \subset (a')^{\perp}$. Let b be any line $\neq z$ in B; then $ab, az, bz \in J$ and $abz \notin J$, hence $abz = 1$ and, similarly, $a'bz = 1$. This yields $a = bz = a'$, a contradiction. Thus our assumption $z \notin A$ is false, and Lemma 6.4 is proved. ∎

LEMMA 6.5. *Let* B *be any 2-Δ-connected point of an S-group plane and* A *a point not connected with* B. *Then* $B^{A^2} = B$.

If A *is also 2-Δ-connected and not orthogonal to* B, *then* $b^{A^2} = B$ *for all* b *in* B.

PROOF. Let a be any line in A and b be any line in B and $q := b^a$. Obviously, we have $q \notin A$, and therefore the mapping $\sigma := \sigma_{qAB}$ is defined. Since B is 2-Δ-connected, we have dom $\sigma = A$, and by Theorem 5.7 the point A^{σ} exists. From $a\sigma = qaa_B = b^a ab = a$ we see that $a \in A^{\sigma}$ and therefore $A^{\sigma} = A$, since A, A^{σ} are two points on a not connected with the 2-Δ-connected point B. If $x \in A$, then $x\sigma, x \in A$, $S(qx)$ and $A \neq S(qx)$ and therefore $x\sigma = x$. This yields $b^{ax} = q^x = x(x\sigma)x_B = x_B \in B$ and thus $b^{A^2} \subset B$. This being true for any line b in B we obtain $B^{A^2} = B$.

If A is also a 2-Δ-connected point not orthogonal to B, then $ab \notin J$ for any lines a, b in A and B, respectively, since $ab \in J$, and Lemma 6.3 implies $A \subset b^{\perp}$ and $B \subset a^{\perp}$ and therefore that A, B are orthogonal. Hence $b^a \notin B$ and to any y in B there is a line x in $A \cap S(b^a y)$. By the first part of our proof $b^{ax} \in B$, and therefore y, $b^{ax} \in B$, $S(b^a y)$ and

$B \neq S(b^a y)$ yields $y = b^{ax}$. Thus we have $B \subset b^{A^2} \subset B$ and therefore $b^{A^2} = B$. ■

LEMMA 6.6. *If a 2-Δ-connected point* A *is not connected with a 1-Δ-connected point* B *in an S-group plane, then there is a line* a *in* A *such that* $B \subset a^\perp$. *If* a *is not a central line, then* $B = a^\perp$.

PROOF. Let b be any line in B. If $A \cap b^\perp = \emptyset$, then we see from Lemma 5.10 that there is a point C on b not connected with A such that

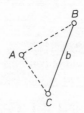

$C^{A^2} = C$. Since A is 2-Δ-connected, we obtain $B = C$. Thus C is Δ-connected, and we deduce from Lemma 5.10 that C is 2-Δ-connected, a contradiction to the fact that B is 1-Δ-connected. Hence $A \cap b^\perp \neq \emptyset$ and there is a line a in $A \cap b^\perp$. Lemma 6.3 yields $B \subset a^\perp$ and Theorem 5.4 yields the last statement of Lemma 6.6. Thus Lemma 6.6 is proved. ■

PROPOSITION 6.7. *Given any S-group plane containing two 2-Δ-connected points not connected by a line. Then*
(a) *There is no central line.*
(b) *There is no polar triangle.*
(c) *Every 2-Δ-connected point is connected with every 1-Δ-connected point.*

PROOF. (a) follows immediately from Lemma 6.4.

For (b): We assume that there is a polar triangle a, b, c. Let A, B be two 2-Δ-connected points not connected by a line. Since A is connected with the triangle a, b, c, we may assume without loss of generality $a \in A$. Moreover we may assume $b \notin A$. Lemma 6.3 yields that A and $S(bc) = a^\perp$ are not connected. Since B is 2-Δ-connected and $A \neq S(ab)$, there is a line d in $B \cap S(ab)$. Clearly, $a \neq d$ and therefore $a^\perp \neq d^\perp$ (otherwise $b \in d^\perp$ and $dac = db \in J$ and therefore $a = d$, a contradiction). If $c \in B$, then $B = a^\perp$, since A is 2-Δ-connected. From Lemma 6.3 we deduce $b^\perp = A = c^\perp$, a contradiction. Thus we have $c \notin B$, and Lemma 6.3 implies that $B \cap d^\perp = \emptyset$. Hence $B \cap a^\perp \neq \emptyset$ and therefore $B = a^\perp$, since A is not

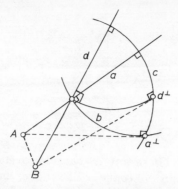

connected with B and with $a^\perp$. Thus $b = d$ and $c \in B$, $d^\perp$, a contradiction. Our assumption there is a polar triangle is not true, and (b) is valid.

For (c): Let A be any 1-Δ-connected point. By our assumption, there are two 2-Δ-connected points B, C not connected by a line. We assume there is a 2-Δ-connected point X not connected with A. Then Lemma 6.6 and (a) imply there is a line a in X such that $A = a^\perp$. Then $a \notin B$ or $a \notin C$. Without loss of generality we may assume $a \notin B$. If $A \cap B = \varnothing$, then by

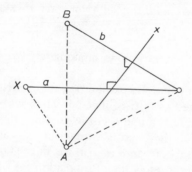

Lemma 6.6 and (a) there is a line b in B such that $A = b^\perp$. Obviously, $a \neq b$ and $S(ab) = x^\perp$ for all x in A, not in $S(ab)$ [we apply (a) and Theorem 5.4]. By (b) A and $x^\perp$ for $x \in A \setminus S(ab)$ are not connected, hence $S(ab) = x^\perp$ for all x in A. Since B, B^x, $x^\perp$ are points on b not connected with A, we obtain $B^x = B$ and $x \notin B$ and therefore $B = x^\perp = S(ab)$, a contradiction to $a \notin B$. Thus there is a line c in $A \cap B$. By Lemma 6.3 $c^\perp$ is a point on a not connected with B. By (b) $A \cap c^\perp = \varnothing$ and therefore $X = c^\perp$ (compare our conclusion just reached). Thus the 2-Δ-connected point X is not connected with the two distinct points A, B on c, a contradiction. Our assumption that there is a 2-Δ-connected point not connected with A is false, and (c) is valid.

Thus Proposition 6.7 is proved. ∎

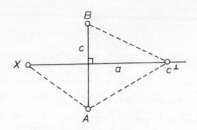

PROPOSITION 6.8. *Given any* Δ-*connected S-group plane containing two* 2-Δ-*connected points not connected by a line, and at least one point, which is completely connected or* 1-Δ-*connected. Then*

(a) *For any* 1-Δ-*connected point A and any line b not in A we have* $A \cap b^\perp \neq \varnothing$.

(b) *For any line a there exists a line b such that* $a \perp b$.

(c) *Any two distinct* 2-Δ-*connected points are not connected.*

(d) *Any line is contained in exactly one* 2-Δ-*connected point.*

(e) *For any line a the set* $a^\perp$ *or the set* $a^\perp \cup \{a\}$ *is a* 2-Δ-*connected point.*

(f) *Any* 2-Δ-*connected point is a set* $a^\perp$ *or a set* $a^\perp \cup \{a\}$ *for some suitable* a.

PROOF.   (a) This statement follows immediately from Lemma 5.10 and Proposition 6.7(c).

For (b): Let a be any line. By assumption there exists a point E, which is completely connected or 1-Δ-connected. Then we may assume $a \notin E$, since if $a \in E$, then there is a line b not in E (since there exist two points not connected by a line there exists a quadrilateral) and we have $a \notin E^b$, and we may interchange E and E^b, or we have $a \in E^b$ and therefore $b \in a^\perp$, as required. But if $a \notin E$, then Proposition 5.11 and (a) imply there is a line $b \in E$, $a^\perp$. This proves (b).

For (c): We assume there are two distinct 2-Δ-connected points A, B connected by a line c. Then (b) yields that there exists a line $a \in c^\perp$. Without loss of generality we may assume $a \notin A$. We deduce from Lemma 6.3, Theorem 5.4, and Proposition 6.7(a) that $c^\perp$ is a point on a not connected with A. We choose any line a' in $c^\perp$ such that $a' \notin B$. Then Lemma 6.3 implies that $c^\perp$, B are not connected. Thus $c^\perp$ is 1-Δ-connected and, by Proposition 6.7(c), connected with the 2-Δ-connected point B, a contradiction. Thus our assumption is false, and (c) is proved.

For (d): Let a be any line. Then there is a 2-Δ-connected point A not on a and, by Lemma 6.3, exactly one point B on a not connected with A. Then B is not completely connected and, by Proposition 6.7(c), not

1-Δ-connected, hence it is 2-Δ-connected. This proves the existence of a 2-Δ-connected point on a, and the uniqueness follows immediately from (c).

For (e): Let a be any line. Let A be the 2-Δ-connected point on a whose unique existence is guaranteed by (d). Then $a^\perp \neq \varnothing$ by (b). If there is a line b in $a^\perp$ not in A, then we see from Lemma 6.3 that $a^\perp$ is a point on b not connected with A and therefore, by Proposition 6.7(c), 2-Δ-connected. If $a^\perp \subset A$, then for $x \in A \setminus \{a\}$ we choose any point $B \neq A$ on x. By (c) the point B is completely connected or 1-Δ-connected, and therefore we can apply Proposition 5.11 and (a) and obtain that there is a line b in $B \cap a^\perp$. From $x, b \in A$, B and $A \neq B$ we deduce $x = b$ and therefore $x \in a^\perp$. This yields $A = a^\perp \cup \{a\}$, and $a^\perp \cup \{a\}$ is a 2-Δ-connected point, as required.

For (f): Let A be any 2-Δ-connected point and b a line in A. Then (b) implies that there exists a line a in $b^\perp$. Obviously, $b \in a^\perp$, $a^\perp \cup \{a\}$, A, and by (e) $a^\perp$ or $a^\perp \cup \{a\}$ is a 2-Δ-connected point. Thus (d) yields $A = a^\perp$ or $A = a^\perp \cup \{a\}$. This proves (f) and concludes the proof of Proposition 6.8. ■

3. THE FIVE TYPES OF COMPLETE S-GROUP PLANES

To separate the complete S-group planes into the classes of the elliptic, Euclidean, Strubecker, hyperbolic-metric, and Minkowskian planes we consider the following three assumptions:

AXIOM H. *There exist two distinct 2-Δ-connected points which are connected by a line.*

AXIOM EM. *There exist two 2-Δ-connected points that are not connected by a line.*

AXIOM $\Delta 1$. *There exists a 1-Δ-connected point.*

The converses of these three axioms are denoted by $\neg$H, $\neg$EM, and $\neg \Delta 1$, respectively. Thus we have six assumptions. These are not independent as the following proposition shows:

PROPOSITION 6.9. *For any Δ-connected S-group plane we have*
(a) *If Axiom H is valid, then EM and $\neg \Delta 1$ are equivalent.*
(b) *If Axiom $\Delta 1$ is valid, then H and $\neg$EM are equivalent.*

PROOF. By Proposition 6.8 we only have to prove: If Axiom $\Delta 1$ and Axiom $\neg$EM are valid, then Axiom H holds. This implication follows immediately from Proposition 6.8 if there exist at least two 2-Δ-connected points. To prove the latter we assume that the complete S-group plane $E(G, S)$ contains at least one 1-Δ-connected point A and at most one 2-Δ-connected point.

If there exists at least one 2-Δ-connected point, say B, then $B^x = B$ for all $x \in S$, and therefore $B = x^\perp$ for all x not in B by Lemma 6.4 and Theorem 5.4. Let a be any line not in B; then Corollary 5.9 yields that there exists a point C on a not connected with B. Then C is 1-Δ-connected, and we can find a line b contained in exactly two points D, E not connected with C. If $b \notin B$, then $B = b^\perp$ and $C \cap b^\perp = \varnothing$, a contradiction to Lemma 5.10. If $b \in B$, then we may assume $D = B$. Obviously, we have $E^a = E$ and $E \subset a^\perp = B = D$, a contradiction.

Thus we are left with the case that there does not exist any 2-Δ-connected point. From Lemma 5.10 we immediately deduce

(1)   *For any line* x *and any point* X *satisfying* $x \notin X$ *we have* $X \cap x^\perp \neq \varnothing$.

Moreover, we show

(2)   *There is no central line.*

Suppose there exists a central line z. Then we can find two points C, D such that $z \in C$ and $C \cap D = \varnothing$. This follows from Corollary 5.9, since there exists the point A which is not completely connected. Let a be any line $\neq z$ in C. Then $a \notin D$, and there is a line b in $D \cap a^\perp$, by (1). Then $z \in S(ab)$ and therefore $b \in C$, D, a contradiction. This proves (2).

From (1), Theorem 5.4, and (2) we deduce that for any line x there is a uniquely determined point containing $x^\perp$. This point may be denoted by S_x. We prove

(3)   *For any line* x *the point* S_x *is completely connected.*

Suppose we can find a line a such that S_a is not completely connected. Then we can find a point C not connected with S_a. Then we have $C \cap a^\perp = \varnothing$ and therefore by (1) $a \in C$. In particular, we obtain $a \notin S_a$, hence $S_a = a^\perp$. Since C is 1-Δ-connected, there exists a line b which is contained in two points D and E, not connected with C. By (1) S_a is connected with D and E, and therefore we obtain without loss of generality, $S_a = D$, since C is Δ-connected. This yields $b \in D = D^a$, E, E^a, and C is not connected with D, E, E^a. Thus $E^a = E$ and $a \notin E$ and therefore $E = a^\perp = S_a = D$, a contradiction. Thus (3) is proved.

(4)   *There is no polar triangle.*

Suppose there exists a polar triangle a, b, c. Then we have $abc = 1$ and $a \in A$ without loss of generality, since A is connected with the triangle a, b, c. Moreover, $a^\perp = S(bc)$ is a completely connected point by (3), and therefore we can find a line b' in $A \cap a^{\perp}$. We put $c' := cbb'$ and obtain $A = S(ab')$ and $ab'c' = 1$. Then $A = (c')^{\perp}$. Therefore A is completely connected by (3), and we arrive at a contradiction. This proves (4).

Now we choose any line a not in A and a point C on a not connected with A (compare Corollary 5.9). Since S_a is completely connected, by (3) we can find a line b in $A \cap S_a$. Then S_b is a completely connected point on a, and we have $S_a \cap S_b \neq \varnothing$. If $S_a \neq S_b$ and $c \in S_a \cap S_b$, then a, b, c would be a polar triangle, which contradicts (4). Thus we have $S_a = S_b = S(ab)$. Let d be any line $\neq a$ in C and $e \in A$, S_d; then our argument yields $S_d = S_e = S(ed)$. Since S_a is completely connected, we can choose a line g in $S_a \cap S_d$. Then $|g^{\perp} \cap S_a| \geq 2$ and $|g^{\perp} \cap S_d| \geq 2$, implying $S_a = S_g = S_d$ and $C = S_g$. However, this contradicts (3), since C is not completely connected.

Thus our assumption must be false, and Proposition 6.9 is proved. ■

Now we define the five types of complete S-group planes as follows: A complete S-group plane $\neq E(G_0, S_0)$, $E(G_1, S_1)$ is called

an *elliptic plane*, if   ($\neg$H), $\neg$EM, $\neg\Delta 1$ holds

a *Euclidean plane*, if $\neg$H, EM, $\neg\Delta 1$ holds

a *Strubecker plane*, if   H, EM, ($\neg\Delta 1$) holds

a *hyperbolic-metric plane*, if   H, ($\neg$EM), $\Delta 1$ holds

a *Minkowskian plane*, if   ($\neg$H), EM, $\Delta 1$ holds

(On the right the assumptions that are bracketed follow from the ones not in brackets.)

The S-group plane $E(G_0, S_0)$ is called a Minkowskian plane, and $E(G_1, S_1)$ is called a hyperbolic-metric plane.

From the definition and from Proposition 6.9 we see that any complete S-group plane is one of the five types enumerated above. Each of the types can be defined also by other assumptions in the frame of the theory of S-groups. Some of the equivalent definitions are noted in the following propositions, though only a few of them are used in the further developments. For characterizations of the five types in relation to other concepts, compare Theorems 6.13, 6.15, 6.21, 6.25, and 6.28.

PROPOSITION 6.10. *For any S-group plane* E *the following are equivalent:*

(a)   E *is an elliptic plane.*

(b)   E *contains only completely connected points and a quadrilateral.*

PROOF. (a) $\Rightarrow$ (b): Let A be any point of an elliptic plane. Then A is not 1-Δ-connected, by $\neg\Delta 1$. If A is 2-Δ-connected, then there is a point B not connected with A. Clearly, B is 2-Δ-connected, and thus EM is valid, a contradiction. Hence A is completely connected. By definition of a complete S-group plane, E contains a quadrilateral.

Obviously, (b) implies (a). Hence Proposition 6.10 is proved. ∎

PROPOSITION 6.11. *For any S-group plane* E *the following are equivalent:*

(a) E *is a Euclidean plane.*

(b)   *Every line of* E *is contained only in completely connected points and in exactly one* 2-Δ-*connected point.*

PROOF. (a) $\Rightarrow$ (b): Let a be any line in a Euclidean plane. By $\neg H$ and $\neg\Delta 1$ all points on a are completely connected with at most one exception. In particular, there exists at least one completely connected point. Hence Proposition 6.8(c) yields that a is contained in exactly one 2-Δ-connected point.

(b) $\Rightarrow$ (a): Let E be any S-group plane satisfying (b). Obviously, E is Δ-connected. There exists at least one line a in E, hence there exists at least one 2-Δ-connected point A and therefore there is a line b not in A. By Lemma 6.3 we can find a point B on b not connected with A. Obviously, B is 2-Δ-connected and therefore EM is valid. Clearly, $\neg\Delta 1$ and $\neg H$ are valid, and there exists a quadrilateral (since there exist two points not connected by a line). Hence E is a Euclidean plane. This proves Proposition 6.11. ∎

PROPOSITION 6.12. *For any S-group plane* $E = E(G, S)$ *the following are equivalent:*

(a) E *is a Strubecker plane.*

(b)   *Every point of* E *is* 2-Δ-*connected.*

(c) E *is a complete S-group plane containing no lines* a, b *such that* a $\perp$ b.

(d) E *is a complete S-group plane and* $Z(G) \cap S^4 \neq \{1\}$.

PROOF. (a) $\Rightarrow$ (b): Let A be any point in a Strubecker plane. If A is completely connected, then Proposition 6.8(b) would imply $\neg H$, a contradiction. Hence A is 2-Δ-connected, and (b) is valid.

(b) $\Rightarrow$ (c): Let E be an S-group plane whose points are all 2-Δ-connected. We assume there are two lines a, b such that $a \perp b$. Since $S(ab)$ is not completely connected, there is a line $c \notin S(ab)$. We put $A := S(ac)$ and obtain from Lemma 6.3 that there is a point B on b not connected with A, since A is 2-Δ-connected. Also B is 2-Δ-connected, hence $A = b^\perp$ by Lemma 6.3 and Proposition 6.7(a). We choose any line $d \neq b$ in B and obtain—by the same argument as before—$S(ad) = b^\perp$ and therefore $A = S(ad)$ and $d \in A, B$, a contradiction. Thus our assumption is false, and (c) is proved.

(c) $\Rightarrow$ (a): Let E be any complete S-group plane without a pair of lines orthogonal to each other. Let A be any point of E. Then there is a line a not in A. Since $a^{\perp} = \varnothing$, we have $A \cap a^{\perp} = \varnothing$ and therefore, by Lemma 5.10, that there is a 2-Δ-connected point B on a not connected with A. We choose any line b in A; then $B \cap b^{\perp} = \varnothing$, and we see from Lemma 5.10 and Lemma 6.3 that A is 2-Δ-connected. Thus all the points of E are 2-Δ-connected and, obviously, H and EM are valid. Thus E is a Strubecker plane.

Up to this point our proof shows that any two of the statements (a), (b), and (c) are equivalent to each other. To prove the equivalence of (d) to the other, we show that (b) implies (d) and that (d) implies (a). For our proof we introduce the following notion: A quadruple a, b, c, d of mutually distinct lines of an S-group plane is called a *dual parallelogram* if $S(ab) \cap S(cd) = \varnothing = S(bc) \cap S(da)$.

(b) $\Rightarrow$ (d): Let $E := E(G, S)$ be any S-group plane whose points are all 2-Δ-connected. First, we show

(1) a, b, c, d *is a dual parallelogram* $\Leftrightarrow$ $abcd \notin S^2$.

If $abcd \notin S^2$, then $S(ab) \cap S(cd) = \varnothing$ and $S(bc) \cap S(da) = \varnothing$ [from $s \in S(ab) \cap S(cd)$ we deduce $abcd = (abs)(scd) \in S^2$ and $s \in S(bc) \cap S(da)$ would imply $abcd = ((bcs)(sda))^a \in S^2$], and therefore a, b, c, d is a dual parallelogram. Conversely, let a, b, c, d be a dual parallelogram. If $abcd = uv$, then $uv \neq 1$ and the 2-Δ-connected point $S(uv)$ is connected with $S(ab)$ or with $S(bc)^a$ by a line t, since $S(ab), S(bc)^a$ are distinct points on b^a. Hence $(tab)(cd) = tuv \in S$ or $(t^a bc)(da) = (tuv)^a \in S$, and therefore $tab \in S(ab) \cap S(cd)$ or $t^a bc \in S(bc) \cap S(da)$, a contradiction. Thus $abcd \notin S^2$.

(2)   *If* a_1, a_2, a_3, a_4 *is a dual parallelogram and* $\alpha = a_1 a_2 a_3 a_4$, *then* $a_i^{\alpha} = a_i$
       for $i = 1, 2, 3, 4$.

By the symmetry in the a_i's, it suffices to prove $a_1^{\alpha} = a_1$. By Lemma 6.5 we have $a_1^{\alpha} = a_1^{a_2(a_3 a_4)} \in S(a_1 a_2)$, since $S(a_1 a_2)$, $S(a_3 a_4)$ are 2-Δ-connected points not connected by a line. Analogously, we obtain $a_1^{\alpha} = a_1^{(a_2 a_3) a_4} \in S(a_4 a_1)$. From a_1^{α}, $a_1 \in S(a_1 a_2), S(a_4 a_1)$ and $S(a_1 a_2) \neq S(a_4 a_1)$ we deduce $a_1^{\alpha} = a_1$, as required.

(3)   *If* a, b, c, d *is a dual parallelogram, then* $\alpha := abcd \in Z(G) \setminus \{1\}$.

Let X be any point; then (1) in the proof of Theorem 5.1 yields that there are lines u, v in X and lines w, z such that $\alpha = uvwz$. By (1) u, v, w, z is a dual parallelogram, and by (2) $u^{\alpha} = u$ and $v^{\alpha} = v$, therefore $X^{\alpha} = X$. This implies $x^{\alpha} = x$ for all lines x and therefore $\alpha \in Z(G)$. Obviously, $\alpha \neq 1$. This proves (3).

(4)   *There exists at least one dual parallelogram.*

Let $S(ab)$ be any point in E. Then $S(ab)$ is a 2-Δ-connected point, and there exists a point $S(cd')$ not connected with $S(ab)$. By Lemma 6.3 we can find a point D on a not connected with $S(bc)$ [clearly, $b \neq c$ by $S(ab) \cap S(cd') = \varnothing$]. Then $S(cd')$, D are connected by a line d, since $S(cd')$ is 2-Δ-connected and not connected with A on a. Then a, b, c, d is a dual parallelogram.

From (4) and (3) we obtain our assumption, that (b) implies (d).

(d) $\Rightarrow$ (a): Let $E := E(G, S)$ be any complete S-group plane such that $Z(G) \cap S^4 \neq \{1\}$, and choose any element $\alpha = abcd \neq 1$ in $Z(G)$. First, we note

(5) $abcd = bcda = cdab = dabc$.

The first equation is clear by $abcd = (abcd)^a = bcda$, and the other cases follow analogously.

Thus we may apply any cyclic permutation to the elements a, b, c, d without affecting the validity of the considered statement. Moreover, we show

(6) $Z(G) \cap S = \varnothing$.

If $z \in Z(G) \cap S$, then $z\alpha \in Z(G) \cap S^5$ and therefore, by the proof of Theorem 5.1 and Theorem 4.4, $z\alpha \in Z(G) \cap S$ or $z\alpha = 1$, hence $z\alpha = z$ by Theorem 4.4(a) and (d). This implies $\alpha = 1$, a contradiction.

By the first part of the proof of (1) and since $\alpha \notin S^2$ [compare Theorem 4.4(b)], we have

(7) a, b, c, d *is a dual parallelogram*.

Furthermore, we prove

(8) $S(ab) \cap c^{\perp} = \varnothing$.

If $b' \in S(ab) \cap c^{\perp}$, we may assume $b' = b$, as otherwise we interchange ab by $a'b'$ for $a' := abb'$. From $(da)^c = b(bcda)c = (bc)(bcda) = da$ we obtain $S(da)^c = S(da)$ and $c \notin S(da)$, hence $(\{b\} \cup S(da)) \subset c^{\perp}$ and $b \notin S(da)$. This implies $c \in Z(G)$ by Theorem 5.4, a contradiction to (6). Thus (8) is valid.

(9) $S(cd)$ *is 2-Δ-connected*.

Since $S(ab) \cap c^{\perp} = \varnothing$, by (8), there is a 2-$\Delta$-connected point A on c such that $c^{S(ab)^2} \subset A$ (compare Lemma 5.10). Hence c, $c^d = c^{cba} = c^{ba} \in A$, $S(cd)$ and $c \neq c^d$ [otherwise $S(bc) \cap d^{\perp} \neq \varnothing$, in contradiction to the statement following from (8) by cyclic permutations of a, b, c, d] and therefore $A = S(cd)$. This proves (9).

By cyclic permutations of the elements a, b, c, d we also obtain that $S(ab)$, $S(bc)$ are 2-Δ-connected. Thus $S(ab)$, $S(bc)$ fulfill Axiom H and $S(ab)$, $S(cd)$ fulfill Axiom EM; therefore E is a Strubecker plane. ∎

The proof of the equivalence of (a) and (d) in Proposition 6.12 almost immediately implies the following interesting:

COROLLARY 6.12′. *Let* (G, S) *be any S-group whose group plane is a Strubecker plane. Then we have for the center of the group* G

$$Z(G) = \{1\} \cup (S^4 \setminus S^2)$$

PROOF. By the Reduction Theorem 5.1, we have $Z(G) \subset S \cup S^2 \cup S^3 \cup S^4$. Since a Strubecker plane contains a quadrilateral (since there is a point that is not completely connected), we deduce from Corollary 4.7 and from Proposition 6.7(a) that $Z(G) \cap (S \cup S^2 \cup S^3) = \{1\}$. Thus we have $Z(G) \subset \{1\} \cup (S^4 \setminus (S \cup S^2 \cup S^3))$. But (1) and (3) in the proof of Proposition 6.12 imply that any element in $S^4 \setminus S^2$ is an element of $Z(G)$; thus we have

$$\{1\} \cup (S^4 \setminus S^2) \subset Z(G) \subset \{1\} \cup (S^4 \setminus (S \cup S^2 \cup S^3)) \subset \{1\} \cup (S^4 \setminus S^2)$$
and therefore $Z(G) = \{1\} \cup (S^4 \setminus S^2)$ ∎

Corollary 6.12′ implies that the center of an S-group whose group plane is a Strubecker plane consists exactly of the dual parallelograms and the unit element.

4. THE ELLIPTIC PLANES

The case of the elliptic planes is the easiest case. By Proposition 6.10 the elliptic planes are exactly those S-group planes that contain a quadrilateral and for which any point is completely connected. Thus not only do any two distinct lines intersect at a point (this is true for any S-group plane, moreover, for any incidence structure), but also any two points are connected by a line. This characterizes the elliptic planes under the S-group planes. More precisely, we wish to prove:

THEOREM 6.13. *For an S-group plane* E *the following are equivalent up to isomorphisms:*
(a) E *is an elliptic plane.*
(b) E *is a projective incidence structure.*
(c) E *is a pappian projective incidence structure.*
(d) E *is an elliptic coordinate plane.*

PROOF. (a) $\Rightarrow$ (b): Let $E(G, S)$ be an elliptic plane and let P denote the set of points in $E(G, S)$. Then Proposition 6.10 yields that the triplet $(P, S, \in)$, consisting of P and S and the inclusion relation $\in$, is a projective plane [for the existence axiom choose any four lines a, b, c, d in S, no three of them concurrent; then $S(ab)$, $S(bc)$, $S(cd)$, $S(da)$ are four points, no three of them collinear]. Hence $E(G, S)$ is a projective incidence structure.

(b) $\Rightarrow$ (c): Let $E(G, S)$ be a projective incidence structure. Then any point in $E(G, S)$ is a completely connected point. From Theorem 5.12 we then deduce that the theorem of Pappus is valid for $E(G, S)$. Hence $E(G, S)$ is a pappian projective incidence structure.

(c) $\Rightarrow$ (d): In this part we construct the quadratic form for an elliptic plane applying Theorem 5.14.

Let $E := E(G, S)$ be an S-group plane that is a pappian projective incidence structure. To apply Theorem 5.14 we verify the assumptions of this theorem.

Since any pappian projective incidence structure is isomorphic to the projective coordinate plane $I(V)$ over a vector space (compare Chapter 2, Section 5, and the Appendix), E is embeddable in a projective coordinate plane $I(V)$. We may identify S and $L(V)$ and, similarly, $\kappa(G, S)$ and $\kappa(V)$.

There exists a completely connected point $S(ab)$ such that $ab \neq ba$, since Corollary 5.3 implies that we can find at most one central line in E. Therefore we can choose two lines a, b such that $b \notin a^{\perp} \cup \{a\}$. Then $S(ab)$ is a completely connected point, since every point is completely connected, and we have $ab \neq ba$.

Theorem 5.14 now implies that there is a quadratic form Q over V such that $(S, \kappa(G, S), \Phi(G, S))$ is isomorphic to an S-subplane of $(L(V, Q), \kappa(V, Q), \Phi(V, Q))$.[*] By our identification we have $L(V) = S \subset L(V, Q) \subset L(V)$ and therefore $L(V, Q) = S = L(V)$. Thus $I(V, Q)$ is a projective incidence structure, and Proposition 2.4 shows that $\text{Ind}(V, Q) = 0$ and $\dim \text{rad}(V, Q) = 0$. Corollary 4.8 yields $\Phi(G, S)$ is injective. Hence $\Phi(V, Q)$ is injective, and therefore we have $\dim V^{\perp} \leq 2$ by Lemma 1.11. By definition $I(V, Q)$ is an elliptic coordinate plane, isomorphic to E. This proves (d).

(d) $\Rightarrow$ (a): Let $I(V, Q)$ be an elliptic coordinate plane. Then Proposition 3.5 shows that $(L(V, Q), \kappa(V, Q), \Phi(V, Q))$ is a complete metric plane. By Theorem 4.9 $I(V, Q)$ is isomorphic to an S-group plane $E(G, S)$. Since $\text{Ind}(V, Q) = 0$ and $\dim \text{rad}(V, Q) = 0$, Proposition 2.4 shows that $I(V, Q)$, and therefore $E(G, S)$ is a projective incidence structure, and that any point of $E(G, S)$ is completely connected. Also there exists a quadrilateral. Hence $E(G, S)$ is an elliptic plane by Proposition 6.10, and

---

[*] We observe that the proof of Theorem 5.14 becomes much easier if we know that the S-group plane $E(G, S)$ is an elliptic plane, and therefore that any point is completely connected.

is isomorphic to $I(V, Q)$. Thus this part is proved, hence Theorem 6.13 is proved. ∎

Theorem 6.13 includes a proof of the Main Theorem 6.1 concerning the elliptic planes and the elliptic coordinate planes. For Theorem 6.2 and the same planes we prove:

THEOREM 6.14. *Let* (G, S) *be an* S-*group whose group plane is an elliptic plane that is isomorphic to the elliptic coordinate plane* $I(V, Q)$ *according Theorem 6.13. Then* (G^*, S^*) *is isomorphic to* $(O(V, Q), S(V, Q))$, *and the center of* G *is equal to* $\{1\}$ *or the cyclic group of order two.*

PROOF. The last statement of Theorem 6.14 follows immediately by Proposition 6.10 and Corollary 5.3.

To prove the other statement we continue the proof of Theorem 6.13, especially the part of the proof showing that (c) implies (d). Let $E(G, S)$ and $I(V, Q)$ be identified and therefore $E(G, S) = I(V)$.

Let a^* be any element in S^*. Then a^* is an axial collineation of $E(G, S)$ which coincides with the collineation $\bar{\sigma}_A$ of $I(V)$ with $a = \langle A \rangle$ (compare the proof of Theorem 5.14). Hence S^* and $S'(V, Q)$ are equal to each other, and therefore the groups G^* and $O'(V, Q)$ coincide. Since there is a bijection μ from $S'(V, Q)$ onto $S(V, Q)$ which induces an isomorphism of $O'(V, Q)$ and $O(V, Q)$ (compare Chapter 3, Section 6), the pairs (G^*, S^*) and $(O(V, Q), S(V, Q))$ are isomorphic. This proves Theorem 6.14. ∎

5. THE EUCLIDEAN PLANES

The next case we wish to treat is the case of Euclidean planes. This is almost as easy as the case of the elliptic planes. We prove:

THEOREM 6.15. *For an* S-*group plane* E *the following are equivalent up to isomorphisms*:
(a) E *is a Euclidean plane.*
(b) E *is an affine incidence structure.*
(c) E *is a pappian affine incidence structure.*
(d) E *is a Euclidean coordinate plane.*

PROOF. (a) $\Rightarrow$ (b): This part is an almost immediate consequence of Proposition 6.11. Let $E = E(G, S)$ be any Euclidean plane and let P

denote the set of all completely connected points in E. Then we consider the triplet $(P, S, \in)$ where $\in$ denotes the inclusion between elements of S and elements of P. For $(P, S, \in)$ we introduce a relation of parallelism in the usual way: Two lines a, b are said to be parallel, denoted by $a \parallel b$, if $a = b$ or if there exists no point in P containing a and b. Then Proposition 6.11 immediately yields:

$$a \parallel b \Leftrightarrow a = b \quad \text{or} \quad S(ab) \quad \text{is 2-}\Delta\text{-connected} \qquad (\circ)$$

The triplet $(P, S, \in)$ is an affine plane (compare the Appendix): Axiom A1 is valid since all the points in P are completely connected and two distinct points of an incidence structure have at most one line in common. Axiom A2 is fulfilled, for if we are given any point A and any line b in S, then Proposition 6.11 yields that b is contained in exactly one 2-Δ-connected point B, and therefore the completely connected point A is uniquely connected with the 2-Δ-connected point B by a line a. Then a is the only line contained in A and parallel to b. To verify Axiom A3, according to Axiom EM we choose any two distinct 2-Δ-connected points $S(ab)$, $S(cd)$ not connected by a line. By Proposition 6.11 the points $S(ac)$, $S(ad)$, $S(bc)$ are three points in P which are not collinear. Thus $(P, S, \in)$ is an affine plane, and therefore $E(G, S)$ is an affine incidence structure. This proves (b).

(b) $\Rightarrow$ (c): Let E denote an S-group plane that is an affine incidence structure. To prove that E is a pappian affine incidence structure we must verify the affine theorem of Pappus. This can be done by applying Theorem 5.12, which involves the projective theorem of Pappus.

All the points of E which are parallel pencils in the sense of the affine incidence structure are 2-Δ-connected, and all the other points are completely connected. Hence (b) in Proposition 6.11 is valid for E, and therefore E is a Euclidean plane.

Now, we wish to prove the affine theorem of Pappus for E. Let A_1, B_2, A_3, B_1, A_2, B_3 be a hexagon and a, b be two lines such that $a \in A_i$ and

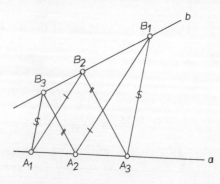

$b \in B_i$ and $a \notin B_i$ and $b \notin A_i$ for $i = 1$, 2, 3. If $(A_1, B_2) \parallel (B_1, A_2)$ and $(B_2, A_3) \parallel (A_2, B_3)$, then we may assume $(A_1, B_2) \neq (B_1, A_2)$ and $(B_2, A_3) \neq (A_2, B_3)$ (otherwise the assertion of the affine theorem of Pappus is trivially valid). We put

$$p_{ik} := (A_i, B_k) \qquad i \neq k; 1 \leq i, k \leq 3$$

By our assumption and ($\circ$) the points $S(p_{12}p_{21})$ and $S(p_{23}p_{32})$ are 2-Δ-connected.

Suppose that the assertion of the affine theorem of Pappus is not true, and therefore $(A_1, B_3) \nparallel (A_3, B_1)$. Then $p_{13} \neq p_{31}$, and the point $S(p_{13}p_{31})$ is not 2-Δ-connected, and thus is a completely connected point. Hence we can find a line c in $S(p_{13}p_{31})$ such that $c \in S(p_{12}p_{21})$. Then we have $c \nparallel p_{23}$, and there exists a line p in $A_3 \cap S(cp_{23})$. We can apply Theorem 5.12 to $P(S(p_{12}p_{21}), B_3, A_3; p_{12}, p_{21}, c; p_{13}, p_{23}, b; a, a, p_{31}, p_{13}, p_{32}, p)$ to obtain $p = p_{32}$. This implies $p_{32} = p_{23}$, since the two parallel lines p_{23} and p_{32} intersect the line c at the same point. However, $p_{23} = p_{32}$ contradicts our previous assumption. Thus our assumption that the affine theorem of Pappus is false has been led to a contradiction. Hence we have verified the affine theorem of Pappus for the Euclidean plane. This proves (c).

(c) $\Rightarrow$ (d): This part of the proof involves the construction of the quadratic form for any Euclidean plane, and is based on Theorem 5.14.*

Let $E = E(G, S)$ be an S-group plane that is a pappian affine incidence structure. For the application of Theorem 5.14 we verify the assumptions of this theorem.

Any pappian affine incidence structure is embeddable in a pappian projective incidence structure, and any pappian projective incidence structure is isomorphic to the projective coordinate incidence structure $I(V)$ (compare the Appendix). Thus we can identify S with a subset of the set $L(V)$ of lines of a suitable projective coordinate incidence structure $I(V)$ such that $L(V) \setminus S$ contains exactly one line w. The concurrency relation $\kappa(G, S)$ can be identified with the restriction of $\kappa(V)$ to S.

Now we wish to show that there exists a completely connected point $S(ab)$ in E such that $ab \neq ba$. As in the preceding part of the proof of our theorem, E is a Euclidean plane. By Proposition 6.11 there is at least one completely connected point A. Let a denote any line through A. From Proposition 6.7 we see that a is not a central line. By Proposition 6.8 we have $A \neq a^\perp \cup \{a\}$, hence there exists a line b in A such that $b \notin a^\perp \cup \{a\}$. Then $A = S(ab)$ is a completely connected point such that $ab \neq ba$.

Thus the assumptions of Theorem 5.14 are fulfilled and, by the

---

* We remark that the proof of Theorem 5.14 becomes much easier if we know that the considered S-group plane is a Euclidean plane.

assertion of this theorem, there is a quadratic form Q on V such that $(S, \kappa(G, S), \Phi(G, S))$ is isomorphic to an S-subplane of the incidence structure $I(V, Q)$ over the metric vector space (V, Q). Since $S \subset L(V, Q)$ and $L(V) \setminus S = \{w\}$, we have $S = L(V, Q)$ or $S \neq L(V, Q)$ and $L(V, Q) = L(V)$. The last case is not possible: If $\langle W \rangle := w \in L(V, Q)$, then we choose linearly independent vectors $A, B \notin w$ such that $A + B \in w$. Then there is a vector C such that $\sigma_W \sigma_A \sigma_B = \sigma_C$ and $C \notin w$ (otherwise A, B are linearly dependent). Hence $\bar{\sigma}_W = \bar{\sigma}_A \bar{\sigma}_B \bar{\sigma}_C$ is the reflection in a line of S, since $\langle A \rangle, \langle B \rangle, \langle C \rangle \in S$, and therefore $w \in S$, a contradiction. Thus $I(V, Q)$ is an affine incidence structure, and therefore we have by Proposition 2.4 that Ind $(V, Q) = 0$ and dim rad $(V, Q) = 1$. Since $\Phi(G, S)$ is injective, by Corollary 4.8 the map $\Phi(V, Q)$ is injective, and therefore we have dim $V^{\perp} \leq 2$ by Lemma 1.11. Hence $I(V, Q)$ is a Euclidean coordinate plane, and (d) is proved.

(d) $\Rightarrow$ (a): Let $I(V, Q)$ be a Euclidean coordinate plane; then Proposition 3.5 yields that $I(V, Q)$ is a complete metric plane that is therefore, by Theorem 4.9, isomorphic to an S-group plane $E(G, S)$. Identifying this S-group plane with $I(V, Q)$, we see that $E(G, S)$ is an affine incidence structure. As shown above, $E(G, S)$ is then a Euclidean plane. Hence $I(V, Q)$ is a Euclidean plane up to isomorphisms.

Thus Theorem 6.15 is proved. ■

Theorem 6.15 yields a proof of the Main Theorem 6.1 in the case of the Euclidean planes and the Euclidean coordinate planes.

THEOREM 6.16. *Let* (G, S) *be an S-group whose group plane is a Euclidean plane that is isomorphic to a coordinate plane* $I(V, Q)$ *according to Theorem* 6.15. *Then* (G^*, S^*) *is isomorphic to the orthogonal group* $(O(V, Q), S(V, Q))$, *and the center* $Z(G)$ *of* G *is equal to* $\{1\}$.

PROOF.   The last statement of Theorem 6.16 follows immediately by Corollary 5.3 and Proposition 6.7.

Now we wish to prove the first statement of Theorem 6.16. We continue the proof of Theorem 6.15. Let $E(G, S)$ be identified with the Euclidean coordinate plane $I(V, Q)$. Then any element a^* in S^* can be embedded in an axial collineation α_a which coincides with the collineation $\bar{\sigma}_A$ of $I(V)$ with $\langle A \rangle = a$ (compare the proof of Theorem 5.14). Hence the sets S^* and $S'(V, Q)$ coincide, and therefore also the groups G^* and $O'(V, Q)$. But this implies that (G^*, S^*) and $(O(V, Q), S(V, Q))$ are isomorphic, since $(O'(V, Q), S'(V, Q))$ and $(O(V, Q), S(V, Q))$ are isomorphic. This concludes the proof of Theorem 6.16. ■

6. THE STRUBECKER PLANES

According to the special role of the Strubecker planes among the well known "classical" cases of metric planes such as the elliptic planes, Euclidean planes, hyperbolic planes, and Minkowskian planes, there are more, and somewhat surprising, characterizations of the Strubecker planes than for the others. The first treatment of the Strubecker planes in the frame of the theory of S-groups was given in the thesis of Ott (compare reference 41). This treatment suggested that the Strubecker plane merely involves concepts belonging to the theory of incidence structures [see Theorem 6.21(d)]. This was very surprising, since the theory of S-groups with Δ-connected points has been developed to describe metric plane geometry. However, the metric concept seemed to disappear in this case, since there exist no two lines orthogonal to each other [compare Proposition 6.12(c)]. Now we know that there exists a very narrow connection to the theory of metric planes: namely, that the Strubecker planes can also be represented as the metric incidence structures of a metric vector space defined by a quadratic form Q (compare Theorem 6.21).

Before formulating theorems for the Strubecker planes which are similar to Theorems 6.13 and 6.14 for the elliptic (and to Theorems 6.15 and 6.16 for the Euclidean) planes, we gather some results in preparation.

PROPOSITION 6.17 (Hoffmann).[*]   *Let* (A, b) *be a flag†* in a pappian *projective incidence structure* (L, κ) *of char* $\neq 2$. *Then the projective reflection group* $(PR(A, b), S(A, b))$ *is an S-group whose group plane is isomorphic to the star-complement in* (L, κ) *with the "star"* A.

PROOF.   We may interpret (L, κ) as the projective coordinate incidence structure $(L(Y), \kappa(V))$ over a field K of char $\neq 2$. Moreover, we may put $V = K^3$. Then the lines of (L, κ) are one-dimensional subspaces of V, and the point A can be regarded as the two-dimensional subspace of V spanned by the vectors $(1, 0, 0)$ and $(0, 1, 0)$. Following the proof of Lemma A2 in the Appendix, we may identify the elements of $S(A, b)$ with the matrices

$$
B(u, v) = \begin{pmatrix} -1 & u & \frac{1}{2}uv \\ 0 & 1 & v \\ 0 & 0 & -1 \end{pmatrix}, \qquad u, v \quad \text{range over } K
$$

---

[*] Compare reference 24.

† We call any pair (X, y) consisting of a point X and a line y in an incidence structure a *flag* if $y \in X$. See the Appendix for this and the definition of projective reflection groups.

and the elements of $P(A, b)$ with the matrices

$$C(w) = \begin{pmatrix} 1 & 0 & w \\ 0 & 1 & 0 \\ 0 & 0 & 1 \end{pmatrix}, \quad w \text{ ranges over } K$$

By Lemma A1 in the Appendix $P(A, b)$ is the center of the group $PR(A, b)$.

From an easy computation we obtain

$$B(u, v)B(u', v')B(u'', v'') = B(u - u' + u'', v - v' + v'')C\left(\frac{1}{2}\begin{vmatrix} u & u' & u'' \\ v & v' & v'' \\ 1 & 1 & 1 \end{vmatrix}\right) \tag{1}$$

Since any matrix $B(x, y)$ is involutory and any matrix $C(w)$ commutes with any matrix $B(x, y)$, and since $C(w)$ is not involutory and is the unit matrix if and only if $w = 0$, we can immediately deduce from equation 1:

$$B(u, v)B(u', v')B(u'', v'') \text{ is involutory} \Leftrightarrow \begin{vmatrix} u & u' & u'' \\ v & v' & v'' \\ 1 & 1 & 1 \end{vmatrix} = 0 \tag{2}$$

Furthermore, a product of a reflection in $S(A, b)$ with an elation $\neq 1$ in $P(A, b)$ is not a reflection in $S(A, b)$ (compare the Appendix). Thus we obtain from equation 1:

$$B(u, v)B(u', v')B(u'', v'') \in S(A, b) \Leftrightarrow \begin{vmatrix} u & u' & u'' \\ v & v' & v'' \\ 1 & 1 & 1 \end{vmatrix} = 0 \tag{3}$$

Finally, we have the following equivalent statements:

The vectors $(u, v, 1)$, $(u', v', 1)$, $(u'', v'', 1)$ are linearly dependent in $K^3 \Leftrightarrow$

the matrix $\begin{pmatrix} u & u' & u'' \\ v & v' & v'' \\ 1 & 1 & 1 \end{pmatrix}$ is not regular $\Leftrightarrow \begin{vmatrix} u & u' & u'' \\ v & v' & v'' \\ 1 & 1 & 1 \end{vmatrix} = 0$

Since every line in $L(V) \backslash A$ is represented by one and only one vector of the form $(u, v, 1)$, the map

$$\psi : \begin{cases} S(A, b) \to L(V) \backslash A \\ B(u, v) \mapsto \langle (u, v, 1) \rangle \end{cases}$$

is a bijective map. From our demonstration above we see for any three elements B_1, B_2, B_3 in $S(A, b)$:

$$B_1B_2B_3 \text{ is involutory} \Leftrightarrow \text{the lines } B_1\psi, B_2\psi, B_3\psi \text{ are concurrent} \quad (4)$$

Since concurrency in $L(V)$ is a ternary equivalence relation, we obtain from equation 4 that the relation

$$B_1B_2B_3 \text{ is involutory}$$

is a ternary equivalence relation in $S(A, b)$. By equations 2 and 3 this relation is equal to the relation

$$B_1B_2B_3 \in S(A, b)$$

This shows that Axiom S is fulfilled for $(PR(A, b), S(A, b))$, hence $(PR(A, b), S(A, b))$ is an S-group. From equation 4 we deduce that ψ is an isomorphism of the group plane of $(PR(A, b), S(A, b))$ onto the star-complement in $(L(V), \kappa(V))$ with the "star" A. Thus Proposition 6.17 is proved. ■

For a further preparation of the "Begründungs" theorem we introduce the ideal plane of a Strubecker plane.

Let $E := E(G, S)$ be any Strubecker plane. Then Proposition 6.12(b) yields that any point of E is 2-Δ-connected. Let P denote the set of all the points of E. Then we establish a relation of parallelism in P: Two points A, B in P are said to be *parallel* if $A = B$ or if $A \neq B$ and $A \cap B = \emptyset$. The relation of parallelism is denoted by $\parallel$. Thus we may write for A, $B \in P$:

$$A \parallel B \Leftrightarrow A = B \quad \text{or} \quad A \neq B \quad \text{and} \quad A \cap B = \emptyset \qquad (\circ)$$

Clearly, the parallelism of points is dual to the parallelism of lines in an affine plane.

Obviously, the relation $\parallel$ in the set P of points is reflexive and symmetric. Moreover, it is transitive: Let $A \parallel B$ and $B \parallel C$. If $A \nparallel C$, then we would have $A \neq C$, and there is a line b such that $b \in A, C$. Hence the 2-Δ-connected B is not connected with the two distinct points A, C on b, a contradiction. Thus $A \parallel C$, and the relation $\parallel$ is an equivalence relation.

Any equivalence class of the relation $\parallel$ in P is called a *parallel pencil of points* or an *ideal line*, and the set of all the ideal lines is called the *point at infinity*, or the *ideal point*, denoted by A_ω.

Interchanging the words "point" and "line" for the Strubecker plane E, we see that the triplet $(S, P, \in)$ is an affine plane, and therefore that the dual group plane $D(G, S)$ is an affine plane. Thus, dualizing the well known procedure of extending an affine plane to a projective plane starting from the Strubecker plane as an incidence structure, we arrive at a projective incidence structure. This incidence structure is called the *ideal incidence structure* of the Strubecker plane $E(G, S)$, denoted by $I(G, S)$. The set L of lines of $I(G, S)$ is the union of S and the set of all the parallel pencils of points. It is clear how the concurrency relation κ of $I(G, S)$ must be defined. The set A_ω is a set of lines in $I(G, S)$ and, moreover, a point of the incidence structure $I(G, S)$. Any point $\neq A_\omega$ in $I(G, S)$ is a set of lines in $I(G, S)$ containing exactly one point A in $E(G, S)$ and exactly one line in A_ω, namely the parallel pencil of points containing A.

Obviously, $L \setminus S$ contains exactly one point; we have $L \setminus S = A_\omega$. Thus $E(G, S)$ is a star-complement in the projective incidence structure $I(G, S)$. Thus we may formulate:

LEMMA 6.18. *Any Strubecker plane* E(G, S) *is a star-complement in its ideal incidence structure* I(G, S).

Now we wish to investigate how the motions of the Strubecker plane $E(G, S)$ can be extended to collineations of the ideal incidence structure $I(G, S)$.

Let α be any element in G. Then the motion α^* of $E(G, S)$ induces a collineation of $I(G, S)$: α^* preserves parallelism of points and therefore maps any parallel pencil of points onto a parallel pencil of points. Thus α^* induces a bijective map of the set of all the ideal lines in $I(G, S)$ onto itself, and therefore α^* induces a bijective map of the set L of all the lines in $I(G, S)$ onto itself. Obviously, concurrent lines in $I(G, S)$ are mapped onto concurrent lines, and nonconcurrent lines onto nonconcurrent lines. Let us denote the collineation of $I(G, S)$ induced by α^* by the same symbol α^*. Then we are allowed to interpret the group G^* as a group of collineations of $I(G, S)$. Any collineation in G^* maps the point A_ω onto itself. Especially, if $\alpha = a \in S$, then the collineation a^* of $I(G, S)$ is the projective reflection in the point A_ω and in the line a, since a^* fixes any point on a and any line through A_ω, and since a^* is involutory by Proposition 6.7(a). Thus the collineation group G^* of $(I(G, S)$ is a projective reflection group (compare the Appendix). Since given any line a in $I(G, S)$ which is not in A_ω, there is a projective reflection in the line a and in the point A_ω, namely a^*, the collineation group G^* of $I(G, S)$ coincides with the group $PR(A)$ (compare the Appendix). Hence the pair (G^*, S^*) for the Strubecker plane $E(G, S)$ may be interpreted as the projective reflection group $(PR(A_\omega), S(A_\omega))$ of the ideal incidence structure $I(G, S)$. We formulate:

LEMMA 6.19. *Let* I(G, S) *denote the ideal incidence structure of the Strubecker plane* E(G, S) *and* A_ω *the point at infinity in* I(G, S). *Then the* *motion group* (G^*, S^*) *of the Strubecker plane* $E(G, S)$ *is isomorphic to the projective reflection group* $(PR(A_\omega), S(A_\omega))$.

To express the parallelism of points in a Strubecker plane $E(G, S)$ in terms of the group G, we prove:

LEMMA 6.20 (Ott).[*] *For distinct lines* a, b *and* c, d *in a Strubecker plane we have*

$$S(ab) \| S(cd) \Leftrightarrow (ab)(cd) = (cd)(ab)$$

PROOF. (A) Let $(ab)(cd) = (cd)(ab)$. If there is a line x such that $x \in S(ab), S(cd)$, then $xab, xcd \in S$. We put $y := xcd$ and obtain $yab = x(cd)(ab) = x(ab)(cd) = xabcd = baxcd = bay$ and therefore $yab = 1$ or $yab \in J$. The first implies $a \perp b$, in contradiction to Proposition 6.12. The second yields $x, y \in S(ab), S(cd)$ and $x \neq y$ (since $c \neq d$), hence $S(ab) = S(cd)$. Thus the points $S(ab), S(cd)$ are either not connected, or identical and therefore parallel.

(B): Let $S(ab), S(cd)$ be parallel. If $S(ab) = S(cd)$, then $(ab)(cd) = (abc)d = c(bad) = (cd)(ab)$. Now, let $S(ab) \neq S(cd)$ and therefore $S(ab) \cap S(cd) = \emptyset$. As in (4) in the proof of Proposition 6.12, we can choose for $S(ab), S(cd)$ a line e in $S(cd)$ such that a, b, c, e is a dual parallelogram. Then (3) in the proof of Proposition 6.12 yields $abce \in Z(G)$, and therefore we can compute $(ab)(cd) = (abce)(ed) = (ed)(abce) = (edc)c(abce) = (cd)(ec)(abce) = (cd)(abce)(ec) = (cd)(ab)$, as required.
This proves Lemma 6.20. ∎

Now we wish to formulate our theorem concerning the "Begründung" for the Strubecker planes as follows (for the dual formulation see Corollary 6.22):

THEOREM 6.21. *For an incidence structure* E *the following are equivalent up to isomorphisms:*
(a) E *is a Strubecker plane.*
(b) E *is an S-group plane that is a star-complement.*
(c) E *is an S-group plane that is a pappian star-complement of char* $\neq 2$.
(d) E *is a pappian star-complement of char* $\neq 2$.
(e) E *is a Strubecker coordinate plane.*

PROOF. We prove the theorem by going along the "circle" (a) $\Rightarrow$ (b) and (b) $\Rightarrow$ (c) and (c) $\Rightarrow$ (d) and (d) $\Rightarrow$ (e) and (e) $\Rightarrow$ (a).

[*] Compare reference 41.

(a) $\Rightarrow$ (b): This follows immediately from Lemma 6.18.

(b) $\Rightarrow$ (c): Let $E = E(G, S)$ be an S-group plane that is a star-complement. Then any point of E is 2-Δ-connected (compare Chapter 2, Section 3), and therefore Proposition 6.12 implies that E is a Strubecker plane.

Let $I(G, S)$ be the ideal incidence structure of the Strubecker plane $E(G, S)$. For any a in S the collineation a^* of $E(G, S)$ induces an involutory homology of $I(G, S)$ by the proof of Lemma 6.19. Thus the projective incidence structure $I(G, S)$ is of char $\neq 2$ if it is pappian (compare the Appendix). Hence we only have to prove that the theorem of Pappus is valid for $I(G, S)$. To prove this it suffices to show that the theorem of Brianchon holds for any configuration with A_ω as its "Pascal-point," since this implies by dualizing that $D(G, S)$ is a pappian affine plane and therefore that the projective extension of $D(G, S)$ is pappian (compare the Appendix). But this projective extension is dual to $I(G, S)$, hence we obtain from this that $I(G, S)$ is pappian. Thus we only have to prove the following statement for the Strubecker plane $E(G, S)$:

(1)   Let a_1, b_2, a_3, b_1, a_2, b_3 *be six lines and* A, B *be two points such that* $a_i \in A$; $a_i \notin B$; $b_i \in B$; $b_i \notin A$ *for* $i = 1$, 2, 3. *If* $S(a_1b_2) \| S(a_2b_1)$ *and* $S(a_1b_3) \| S(a_3b_1)$, *then* $S(a_2b_3) \| S(a_3b_2)$.

We prove (1) as follows[*]: From $a_i \in A$ and $b_i \in B$ for $i = 1$, 2, 3 we deduce $a_1a_2a_3$, $b_1b_2b_3 \in S$. By Lemma 6.20 $S(a_1b_2) \| S(a_2b_1)$ yields $(a_1b_2)(a_2b_1) = (a_2b_1)(a_1b_2)$ and $S(a_1b_3) \| S(a_3b_1)$ implies $(a_1b_3)(a_3b_1) = (a_3b_1)(a_1b_3)$. Thus we can compute

$$
\begin{aligned}
(a_2b_3)(a_3b_2) &= (a_2a_1)(a_1b_3)(a_3b_2) = (a_2a_1)(a_3b_1a_1b_3b_1a_3)(a_3b_2) \\
&= (a_2a_1a_3)(b_1a_1)(b_3b_1b_2) = (a_3a_1)(a_2b_1)(a_1b_2)(b_1b_3) \\
&= (a_3a_1)(a_1b_2)(a_2b_1)(b_1b_3) = (a_3b_2)(a_2b_3),
\end{aligned}
$$

hence $S(a_2b_3) \| S(a_3b_2)$ by Lemma 6.20. This proves (1), and therefore (c) holds.

(c) $\Rightarrow$ (d) follows trivially.

(d) $\Rightarrow$ (e): Let $E = (L, \kappa)$ denote a pappian star-complement of char $\neq 2$. Then there is a pappian projective incidence structure (L', κ') such that (L, κ) is a substructure and $L \setminus L'$ is a point A. By the Appendix we may assume that (L', κ') is the projective coordinate incidence structure over a field K of char $\neq 2$. Then we may assume without loss of generality that $V = K^3$ and $A = \langle (1, 0, 0), (0, 1, 0) \rangle$. We consider the map

$$
Q : \begin{cases} V \to K \\ (x, y, z) \mapsto z^2 \end{cases}
$$

---

[*] This proof is due to U. Ott [41].

Obviously, Q is a quadratic form, and therefore (V, Q) is a metric vector space. Moreover, we have $\operatorname{rad}(V, Q) = A$ and therefore $\dim \operatorname{rad}(V, Q) = 2$. Since every isotropic vector is in $\operatorname{rad}(V, Q)$, Lemma 1.6 yields $\operatorname{Ind}(V, Q) = 0$. Moreover, we have $V^{\perp} = A$ and therefore $\dim V^{\perp} = 2$. By definition $I(V, Q)$ is a Strubecker coordinate plane. Obviously, $I(V, Q)$ is isomorphic to (L, κ). This proves (e).

(e) $\Rightarrow$ (a): Let $I(V, Q)$ denote a Strubecker coordinate plane. Then we have for the metric vector space (V, Q) over the field K $\dim \operatorname{rad}(V, Q) = 2$ and $\dim V^{\perp} \leq 2$, hence $\dim V^{\perp} = 2$. This implies $\operatorname{char} K \neq 2$ by Lemma 1.8. Thus $I(V)$ is a pappian projective incidence structure of $\operatorname{char} \neq 2$. Let A be the point $\operatorname{rad}(V, Q)$ in $I(V)$ and b be any line in A. Then we deduce from Proposition 6.17 that the projective reflection group $(PR(A, b), S(A, b))$ is an S-group whose group plane is isomorphic to the star-complement in $(L(V), \kappa(V))$ with the "star" A. But this star-complement is equal to $I(V, Q)$. As proved above, any S-group plane that is a star-complement is a Strubecker plane. Thus the S-group plane $E(PR(A, b), S(A, b))$ is a Strubecker plane and is isomorphic to $I(V, Q)$. This proves (a) and completes the proof of Theorem 6.21. ■

Dualizing the statements (b) and (d) in Theorem 6.21, we obtain:

COROLLARY 6.22 (Ott).[*] *For an affine plane* A *the following are equivalent up to isomorphisms:*

(a) A *is the dual group plane of an S-group.*

(b) A *is a pappian affine plane of char $\neq 2$.*

Next we wish to prepare the analogous theorem to Theorem 6.14 and Theorem 6.16 by the following statement:

PROPOSITION 6.23 (Ott).[*] *Let* E(G, S) *and* E(G', S') *be two isomorphic Strubecker planes. Then* (G, S) *and* (G', S') *are isomorphic.*

PROOF. If $E(G, S)$ and $E(G', S')$ are isomorphic Strubecker planes, then by Theorem 6.21 $E(G, S)$ and $E(G', S')$ are isomorphic pappian star-complements of $\operatorname{char} \neq 2$. Let Φ denote the isomorphism of $E(G, S)$ onto $E(G', S')$. Then $S\Phi = S'$. Dualizing the well known result that an isomorphism of two affine planes can be extended to an isomorphism of the projective extensions of these affine planes (compare the Appendix), we may extend Φ to an isomorphism of $I(G, S)$ onto $I(G', S')$. This isomorphism is also denoted by Φ.

[*] Compare reference 41.

As shown above (in proving Lemma 6.19) we may interpret a^* for a in S as a projective reflection of $I(G, S)$ and $(a')^*$ for a' in S' as a projective reflection of $I(G', S')$. The center of a^* is the point at infinity A_ω of $I(G, S)$, and the center of $(a')^*$ is the point at infinity $A_{\omega'}$ of $I(G', S')$. Obviously, Φ maps A_ω onto $A_{\omega'}$. By the uniqueness of the projective reflection in a given point and a given line (see the Appendix), we have

$$(a\Phi)^* = \Phi^{-1}a^*\Phi \qquad \forall a \in S \tag{1}$$

Trivially, we have for a, b in S

$$ab = 1 \Leftrightarrow (a\Phi)(b\Phi) = 1 \tag{2}$$

Moreover, for a, b, c, d in S we prove

$$abcd = 1 \Leftrightarrow (a\Phi)(b\Phi)(c\Phi)(d\Phi) = 1 \tag{3}$$

To prove this we show the following statement for any four lines a, b, c, d in an arbitrary S-group plane that contains a quadrilateral:

$$abcd = 1 \Leftrightarrow a, b, c \quad \text{are concurrent and} \quad a^*b^*c^*d^* = 1 \tag{4}$$

Obviously, $abcd = 1$ implies a, b, c are concurrent and $a^*b^*c^*d^* = 1$. Conversely, let a, b, c be concurrent and $a^*b^*c^*d^* = 1$; then there is a line e such that $abc = e$, and we obtain $e^* = a^*b^*c^* = d^*$ and therefore $x^{ed} = x$ for all x in S. This yields $ed \in Z(G)$, and by Theorem 4.4(b) $ed = 1$ and thus $abcd = abce = 1$.

Now we prove equation 3 by applying equations 1 and 4: $abcd = 1 \Leftrightarrow a$, b, c are concurrent and $a^*b^*c^*d^* = 1 \Leftrightarrow a\Phi$, $b\Phi$, $c\Phi$ are concurrent and $(a\Phi)^*(b\Phi)^*(c\Phi)^*(d\Phi)^* = \Phi^{-1}(a^*b^*c^*d^*)\Phi = 1 \Leftrightarrow (a\Phi)(b\Phi)(c\Phi)(d\Phi) = 1$.

Statement 3 can be extended as follows (for $a_1, a_2, \ldots, a_n \in S$):

$$a_1 a_2 \cdots a_n = 1 \Leftrightarrow (a_1\Phi)(a_2\Phi) \cdots (a_n\Phi) = 1 \tag{5}$$

Obviously, $n \neq 1$, and by Proposition 6.7(b), $n \neq 3$. Equation 5 follows from equations 2 and 3 if $n = 2$ and $n = 4$. Now let $n \geq 5$. Following the proof of the Reduction Theorem 5.1, we can reduce $a_1 a_2 \cdots a_n = 1$ by using equations of the form $abcd = 1$ to a product of at most four elements of S. Thus equation 3 yields equation 5, since in any step of this reduction process we can apply equations 3 and 2, and the procedure ends by a relation as in equation 2 or 3.

By equation 5 and the fact that any element α in G is the product of elements in S, we can extend the map Φ of S onto S' to a map $\bar{\Phi}$ of G onto G' by putting for $\alpha = a_1 \cdots a_n$ in G and $a_1, \ldots, a_n \in S$:

$$\alpha\bar{\Phi} = (a_1 \cdots a_n)\bar{\Phi} := (a_1\Phi) \cdots (a_n\Phi)$$

The definition of $\bar{\Phi}$ does not depend on the representation of α as a product of elements in S, since $a_1 \cdots a_n = a_1' \cdots a_m'$ implies $a_1 \cdots a_n a_m' \cdots a_1' = 1$, and therefore by equation (5), $(a_1\Phi) \cdots (a_n\Phi) \times (a_m'\Phi) \cdots (a_1'\Phi) = 1$, hence $(a_1 \cdots a_n)\bar{\Phi} = (a_1' \cdots a_m')\bar{\Phi}$. Obviously, the map

$$\bar{\Phi} : \begin{cases} G \to G' \\ a_1 a_2 \cdots a_n \mapsto (a_1\Phi)(a_2\Phi) \cdots (a_n\Phi) \end{cases}$$

is a homomorphism of G into G'. By equation 5, $\bar{\Phi}$ is injective. Since $S\Phi = S'$, the map $\bar{\Phi}$ is surjective and thus an isomorphism of the S-group (G, S) onto the S-group (G', S'). This proves Proposition 6.23. ∎

The analogous theorem to Theorems 6.14 and 6.16 can be formulated as follows:

THEOREM 6.24. *Let* (G, S) *be any S-group whose group plane is isomorphic to a Strubecker coordinate plane* I(V, Q) *over the field* K. *Then*

(a) (G, S) *is isomorphic to the projective reflection group* $(PR(A, b), S(A, b))$ *for any flag* (A, b) *of the projective coordinate incidence structure* I(V).

(b) (G^*, S^*) *is isomorphic to the projective reflection group* $(PR(A), S(A))$ *for any point* A *of* I(V).

(c) (G^*, S^*) *is isomorphic to the orthogonal group* $(O(V, Q), S(V, Q))$.

(d) *The center* $Z(G)$ *of the group* G *is isomorphic to the additive group* K^+ *of the field* K.

PROOF. Let (G, S) be any S-group whose group plane is isomorphic to a Strubecker coordinate plane $I(V, Q)$ over a field K. Then $E(G, S)$ is itself a Strubecker plane (compare Theorem 6.21). Moreover, Theorem 6.21 yields that the ideal incidence structure $I(G, S)$ is a pappian projective incidence structure of char $\neq 2$, and that $E(G, S)$ is a star-complement in $I(G, S)$ with the star A_ω, the point at infinity in $I(G, S)$. Obviously, the isomorphism of $E(G, S)$ and $I(V, Q)$ can be extended to an isomorphism Φ of $I(G, S)$ and $I(V)$.

For (a): Let (A, b) be any flag in $I(V)$. Since the projective reflection group $(PR(A, b), S(A, b))$ in the pappian projective incidence structure $I(V)$ is isomorphic to the projective reflection group $(PR(A', b'), S(A', b'))$ for any flag (A', b') (compare the Appendix), we may assume without loss of generality that $A = A_\omega \Phi$. Then $A = \text{rad}(V, Q)$. By Proposition 6.17 $(PR(A, b), S(A, b))$ is an S-group whose group plane is isomorphic to the star-complement in $I(V)$ with the "star" A. Hence $E(PR(A, b), S(A, b))$ is isomorphic to $I(V, Q)$ and therefore to

$E(G, S)$. Proposition 6.23 then yields that $(PR(A, b), S(A, b))$ and (G, S) are isomorphic.

For (b): This follows immediately from (a) and Lemma A1 and A2 in the Appendix (compare also Lemma 6.19).

For (c): We put $A := A_\omega \Phi$. Then $A = \operatorname{rad}(V, Q)$. Let $\langle X \rangle$ be any line in $I(V)$ not in the point A in $I(V)$. Then $\langle X \rangle \in L(V, Q)$ and σ_X is defined. Moreover, we have $\langle X \rangle^\perp = \operatorname{rad}(V, Q)$, since $\langle X \rangle^\perp \supset V^\perp \supset \operatorname{rad}(V, Q)$ and $\dim \operatorname{rad}(V, Q) = 2$ and $\dim \langle X \rangle^\perp = 2$ (the last by Lemma 1.1 and the fact that $\operatorname{char} K \neq 2$ implies $g(X, X) = 2Q(X) \neq 0$ and therefore $X \notin V^\perp$). Hence the collineation $\bar{\sigma}_X$ of $I(V)$ induced by σ_X is a projective reflection in the point A and in the line $\langle X \rangle$, thus an element of $S(A)$. Thus the set $S'(V, Q)$ of all the collineations $\bar{\sigma}_X$ such that $\sigma_X \in S(V, Q)$ is contained in $S(A)$. On the other hand, any element σ in $S(A)$ is a projective reflection in the point A and in a line x not in A. If $\langle X \rangle := x$, then σ_X is defined and $\bar{\sigma}_X$ coincides with σ. Hence $S'(V, Q) = S(A)$ and therefore $(O'(V, Q), S'(V, Q)) = (PR(A), S(A))$. Since $(O'(V, Q), S'(V, Q))$ and $(O(V, Q), S(V, Q))$ are isomorphic, (c) follows from (b).

For (d): Since the center of the S-group $(PR(A, b), S(A, b))$ for any flag (A, b) in $I(V)$ is the group $P(A, b)$, and since $P(A, b)$ is isomorphic to the additive group K^+ of the field K (compare the Appendix), statement (d) follows immediately from (a).

Thus Theorem 6.24 is proved. ∎

Theorems 6.21 and 6.24 contain a complete proof of Theorems 6.1 and 6.2 concerning the case of Strubecker planes.

7. THE HYPERBOLIC-METRIC PLANES

We wish to treat the case of hyperbolic-metric planes quite differently from the cases of plane metric geometry considered in the preceding sections. Our investigations are based on the famous "Endenrechnung" introduced in Hyperbolic Geometry by D. Hilbert, and are generalized and formulated purely in terms of the calculus of reflections by F. Bachmann [2]. Our treatment closely follows Bachmann [2], §11.

Now let $E(G, S)$ denote any hyperbolic-metric plane distinct from $E(G_1, S_1)$. Then we call any 2-Δ-connected point in $E(G, S)$ an *end*. By Axiom H there are two distinct ends. Proposition 6.8(c) then yields

(1)   *Any two ends are connected by a line.*

If there is a central line z, then we see by Lemma 6.4 that any end is incident with z. We furthermore prove

(2)   *No end* A *is contained in a set* $b^\perp$.

Suppose there is an end A and a line b such that $A \subset b^\perp$. Then there is no central line z, as otherwise we would have $z \in A$ and therefore $zab = 1$ for any $a \ne z$ in A, a contradiction to Corollary 4.6. By Axiom H there is an end $B \ne A$ and from (1) we see that there is a line $s \in A \cap B$. If $b \in B$, then we choose any line $t \ne s$ in A and deduce from Lemma 6.3 that $B \cap b^\perp = \varnothing$ and therefore $A \cap B = \varnothing$, a contradiction. Thus we have $b \notin B$. Since $s \perp b$, we deduce from Lemma 6.3 that A, B are not connected with the point $D := s^\perp$ on b. Since $A^b = A$, we have $B^b = B$, and therefore $B \subset b^\perp$ and b would be a central line, a contradiction. This proves (2).

Now (2) implies immediately

(3)   *If* A *is an end, then*

$$A^x = A \Leftrightarrow x \in A$$

To prepare the "Endenrechnung" we choose any end O and put

$$T := \{O^\alpha \mid \alpha \in G\}$$

Thus T is an orbit of the point O under the group G. Since T only consists of ends, any two points in T are connected by a line.

(4)   *For all* A *in* T *we have* $A^{S^2} = A^S$.

Since $A^S = A^{aS} \subset A^{S^2}$ for $a \in A$, it suffices to prove that $A^{S^2} \subset A^S$. Let $B := A^{ab}$ for $ab \in S^2$. If $A = B$, then $A^s = B$ for all s in A. Suppose $A \ne B$; then $a \ne b$ and $A \cap S(ab) \ne \varnothing$, as otherwise $B = A^{ab} = A$ by Lemma 6.5. Let $s \in A \cap S(ab)$; then $A^{(sab)} = B$ and $sab \in S$.
   From (1) we immediately deduce

(5)   *For all* A *in* T *we have* $T = A^S$.

Let A, B denote two ends in T. Then a point C such that

$$x \in C \Leftrightarrow A^x = B \qquad (*)$$

is denoted by T_{AB}. Obviously, T_{AB} is uniquely determined by the ends A, B. Moreover, if $A \ne B$, then A, B are connected and we have

$T_{AB} \subset (A, B)^{\perp}$. We show

(6) *Given any two ends* A, B *in* T, *there exists exactly one point* T_{AB}. *If* A = B, *then* T_{AB} = A, *and if* A ≠ B, *then* T_{AB} *is* 1-Δ-*connected and not connected with* A *and* B.

If $A = B \in T$, then we see from (3) that equation (∗) is valid for $C := A$. Hence we may put $T_{AA} := A$. If $A \neq B$, then by (5) there is a line s such that $A^s = B$, and by (1) there is a line a in $A \cap B$. Then $a \perp s$, and there is exactly one point D on s not connected with A and such that $D \subset a^{\perp}$ (compare Lemma 6.3). Obviously, $D = D^s$ and $B = A^s$ are not connected. By Lemma 6.5, we obtain $A^x = (A^{D^2})^x = (A^{sD})^x = B^{D^2} = B$ for all x in D. Conversely, let $A^y = B$ for any element y in S. Then $y \notin A$, B, and therefore $y \neq a$ and D is connected with $S(ay)$ by a line x. Then $A^{ayx} = A^{yx} = B^x = A$ and $ayx \in S$, and therefore by (3) $ayx \in A$. This implies $a, ayx \in A$, $S(ay)$ and $A \neq S(ay)$, hence $a = ayx$ and $y = x \in D$. Thus we may put $T_{AB} := D$. Since D is not connected with the two distinct points A, B on a, we see that A is 1-Δ-connected.

(7) *Given* A, B, C, D *in* T *and* A, B ≠ C, D *there is exactly one* s *in* S *such that* A^s = B *and* C^s = D.

To prove (7) it suffices to show that T_{AB} and T_{CD} are distinct and connected by a line.

If $A = B$ and $C = D$, then (6) implies $T_{AB} = T_{AA} = A \neq C = T_{CC} = T_{CD}$ and T_{AB}, T_{CD} are connected by (1).

If $A \neq B$ and $C = D$, then A, B are connected by (1), and T_{AB} is not connected with A, B by (6). However, we see from (1) that A, B are connected with C. Thus $T_{AB} \neq C$, and by (6) $C = T_{CD}$, thus $T_{AB} \neq T_{CD}$. Since our assumption yields that A, B, C are mutually distinct, the Δ-connected point T_{AB} is connected with one of these points, hence connected with T_{CD}, since $T_{AB} \cap A = \varnothing = T_{AB} \cap B$.

If $A \neq B$ and $C \neq D$, then T_{CD} is not connected with C, D, but—see our argument above—is connected with A and B. Hence $T_{AB} \neq T_{CD}$. Moreover, A, B, T_{CD} are mutually distinct (since T_{CD} is not 2-Δ-connected and therefore not an end). Therefore the Δ-connected point T_{AB} is connected with T_{CD}, since $T_{AB} \cap A = \varnothing = T_{AB} \cap B$.

(8) T *contains at least three ends.*

Since O is not completely connected, there is a line a not in O. From (3) follows $O^a \neq O$. We choose any line b in O^a not in O. Since (6) yields

$T_{OO^a} \cap O^a = \varnothing$, we obtain $b \notin T_{OO^a}$ and therefore $O^a \neq O^b$. From (3) we see that $O^b \neq O$. Thus O, O^a, O^b are three mutually distinct ends in T.

(9) *We have* $S^4 \subset S^2$.

Suppose that there are lines a, b, c, d such that $abcd \notin S^2$. Then $a \neq b$ and $c \neq d$ and $S(ab) \cap S(cd) = \varnothing$ [from $s \in S(ab) \cap S(cd)$ we deduce $abcd = (abs)(scd) \in S^2$]. By (1) in the proof of the Reduction Theorem 5.1, we may assume $O = S(ab)$. Moreover, we may assume $c, d \in a^\perp$, since $O^c \neq O$ and $O^{cd} = O$ by (3) and Lemma 6.5, hence for $a' \in O$, $O^c = O^d$, then $c, d \in a'^\perp$, and we may interchange ab by $a'b'$ with $b' := a'ab$. From $O \notin c^\perp$, by (2) we see that $c^b \neq c$ and therefore $S(dc) \neq S(dc^b)$ and that the 2-Δ-connected point O is connected with $S(dc^b)$ by a line s. Then we have $abcd = (dabc)^d = (badbcb)^{bd} = (bas)^{bd}(sdc^b)^{bd} \in S^2$, in contradiction to our assumption. This proves (9).

(10) *If there exists a line z such that $z \in X$ for all X in T, then $z \in Z(G)$.*

By (5) $z \in X$ for all X in T implies $z \in O^s$ for all s in S and therefore $s \in z^\perp$ for all s not in O, hence $z \in Z(G)$ by Theorem 5.4.

(11) *There exists a central line or a polar triangle.*

Suppose that no central line exists. By (8) and (10) we can find three mutually distinct ends A, B, C in T which are not collinear. Then B, C are connected by a line a, and A, A^a are connected by a line b. Since $a \notin A$ and $a \perp b$, Lemma 6.3 implies that $b^\perp$ is a point on a by our assumption that there is no central line. By (2), we have B, $C \neq b^\perp$. Hence T_{BC} is connected with $b^\perp$ by a line c [compare (6)]. Since $T_{BC} \subset a^\perp$, we obtain $abc = 1$.

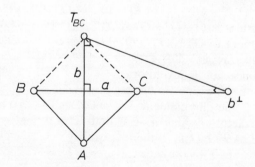

(12) *If a polar triangle exists, then* $G = S^2$; *and if there exists a central line* z, *then* $G = S^2 \cup (zS^2)$.

By the Reduction Theorem 5.1 and by (9) we have $G = S^2 \cup S^3$ (since $S \subset S^3$). If there is a polar triangle a, b, c, we deduce from (9) $S^3 = abcS^3 \subset S^6 \subset S^2$ and therefore $G = S^2$. If there is a central line z, then (9) implies $zS^3 \subset S^4 \subset S^2$, and therefore $S^3 \subset zS^2$ and $G = S^2 \cup (zS^2)$, as required.

(13)   *If α^* for $\alpha \in G$ fixes three mutually distinct ends in* T, *then $\alpha^* = 1$.*

From (12) we see that it is sufficient to consider the case $\alpha = ab$. Let A, B, C denote three mutually distinct ends in T fixed by α^*. If $\alpha \neq 1$, then $S(\alpha) = T_{AA^a} = T_{BB^a} = T_{CC^a}$ by (6). However, this yields $\{A, A^a\} = \{B, B^a\} = \{C, C^a\}$, and A, B, C cannot be mutually distinct, a contradiction. Hence $\alpha = 1$ and therefore $\alpha^* = 1$.

Now, we wish to develop the "Endenrechnung." We choose any end $U \neq O$ in T (the "point at infinity") and denote the line connecting O and U by o. We call O the zero element and put

$$K := \{O^u \mid u \in U\}$$

Then $K \cup \{U\} = T$, since if A is an end $\neq U$ in T, then (7) yields that to U, $U \neq O$, A there is an element u such that $U^u = U$ and $O^u = A$. By (3) we have $u \in U$.

We define an addition in K by

$$O^u + O^v := O^{uov} \tag{$\circ$}$$

Since u, o, $v \in U$, we have $uov \in U$ and therefore $O^{uov} \in K$.

(14) $(K, +)$ *is an abelian group.*

The associative law is fulfilled: $(O^u + O^v) + O^w = O^{uov} + O^w = O^{(uov)ow} = O^{uo(vow)} = O^u + O^{vow} = O^u + (O^v + O^w)$ for all O^u, O^v, O^w in K.
The neutral element is O^o : $O^o + O^u = O^{oou} = O^u$ for all O^u in K.
The inverse element to O^u is $O^{(uo)}$: $O^u + O^{(uo)} = O^{uouo} = O^{uoouo} = O^o = O$.
The commutative law is valid: $O^u + O^v = O^{uov} = O^{vou} = O^v + O^u$ for all O^u, O^v in K, since uov is involutory.

To define a multiplication we choose any end $E \neq O$, U in T; it exists by (8). We call E the unit element. We put $Z := T_{OU}$. To any end $A \neq O$ in K there is exactly one line a such that $E^a = A$ and $a \in Z$, since (7) yields that to E, $A \neq O$, U in T there is exactly one element a such that $E^a = A$ and $O^a = U$. Obviously, $a \in T_{OU} = Z$. Thus we have

$$K \setminus \{O\} = \{E^a \mid a \in Z\}$$

Let e be the line in Z such that $E^e = E$. Then we define a multiplication in K by

$$\begin{cases} E^a \cdot E^b = E^{aeb} \\ O \cdot A = A \cdot O = O \qquad \text{for all } A \text{ in } K \end{cases} \tag{$\circ\circ$}$$

Obviously, $E^a \cdot E^b \in K$, since $a, e, b \in Z$ and therefore $aeb \in Z$. We show

(15) $(K \setminus \{O\}, \cdot)$ *is an abelian group.*

The associative and commutative law follow as in the proof of (14). $E^e = E$ is the unit element: $E^e \cdot E^a = E^{eea} = E^a$ for all E^a in $K \setminus \{O\}$. The end $E^{(a^e)}$ is inverse to E^a, since $E^{ae(a^e)} = E^{aeeae} = E$. This proves (15).

Now we wish to prove one of the distributive laws. We prepare with

(a) $O^u \cdot E^a = O^{uea}$ $\qquad \forall u \in U; \forall a \in Z$

This follows if $u = o$ from $O \cdot E^a = O = O^{aeoea}$. If $u \neq o$, then there is an element b in Z such that $O^u = E^b$, and we can compute $O^u \cdot E^a = E^b \cdot E^a = E^{bea} = (O^u)^{ea} = O^{aeuea} = O^{uea}$, as required. From (a) we deduce a distributive law: $(O^u + O^v) \cdot E^a = O^{uov} \cdot E^a = O^{(uov)ea} = O^{ueaovea} = O^{uea} + O^{vea} = O^u \cdot E^a + O^v \cdot E^a$ for all O^u, O^v in K and for all E^a in $K \setminus \{O\}$. Trivially, we have $(O^u + O^v) \cdot O = O^u \cdot O + O^v \cdot O$ for all O^u, O^v in K. Together with (14) and (15), this proves

(16) $(K, +, \cdot)$ *is a field.*

This field is called a *field of ends* in the hyperbolic-metric plane.

The next step is to construct a three-dimensional metric vector space (V, Q) over a field K of ends in the hyperbolic-metric plane $E(G, S)$ such that $(G^*, S^*) \cong (O(V, Q), S(V, Q))$. In preparation we define, in our notations introduced above,

$$S' := U \cup Z$$

and prove

(17) S' *generates the group* G.

It suffices to prove that S' generates S. Let s denote any element in $S \setminus S'$; then $s \notin U$ and by (3) $U^s \neq U$. Hence there is an element u in U such that $O^u = U^s$. This yields $O^{usu} = U$, and therefore by (6) $a := usu \in T_{OU} = Z$. Thus $s = a^u$ and $u \in U$ and $a \in Z$.

If $\alpha \in G$, then we denote the restriction of α^* to the set T of ends by $\bar{\alpha}^*$. For $N^* \subset G^*$ let $\bar{N}^* := \{\bar{\alpha}^* \mid \alpha^* \in N^*\}$. Then

$$\delta : \begin{cases} G^* \to \bar{G}^* \\ \alpha^* \mapsto \bar{\alpha}^* \end{cases}$$

is a homomorphism. From (13) we see that $\ker \delta = \{1\}$, and therefore that δ is an isomorphism. We note

(18)   *The map δ is an isomorphism of G^* onto $\bar{G}^*$.*

Now let V denote the three-dimensional vector space over the field K of ends considered above, whose elements are the 2×2-matrices

$$\mathbf{A} = \begin{pmatrix} A_1 & A_2 \\ A_3 & -A_1 \end{pmatrix}, \qquad A_1, A_2, A_3 \in K$$

Moreover, let Q be the map

$$Q : \begin{cases} V \to K \\ \mathbf{A} \mapsto -A_1{}^2 - A_2 A_3 \end{cases}$$

Then Q is a quadratic form and (V, Q) a metric vector space over the field K of ends. We have $\dim V^\perp \leq 1$ and $\mathrm{Ind}\,(V, Q) = 1$ and $\dim \mathrm{rad}\,(V, Q) = 0$ (by Lemmas 1.7 and 1.6). If we denote by $\mathbf{E}$ the unit matrix and by $\mathbf{AB}$ the product of the matrices $\mathbf{A}$, $\mathbf{B}$ in V, then we have

(b) $Q(\mathbf{A}) = det\ \mathbf{A}$.

(c) $g_Q(\mathbf{A}, \mathbf{B})\mathbf{E} = -(\mathbf{AB} + \mathbf{BA})$.

(d) $Q(\mathbf{A})\mathbf{E} = -\mathbf{A}^2$.

(e) $\mathbf{A}$ *is isotropic* $\Leftrightarrow$ *det* $\mathbf{A} = O$.

(f)   *If $\mathbf{A}$ is nonisotropic, then* $\mathbf{A}^{-1} = -\mathbf{A}/Q(\mathbf{A})$.

Moreover, obviously we have

$\langle \mathbf{A} \rangle \in L(V) \setminus L(V, Q) \Leftrightarrow \mathbf{A}$ is a multiple of the following matrices:

$$\begin{pmatrix} E & A^{-1} \\ -A & -E \end{pmatrix} \quad \text{for } A \neq O \text{ in } K \quad \text{or} \quad \begin{pmatrix} O & O \\ E & O \end{pmatrix} \quad \text{or} \quad \begin{pmatrix} O & E \\ O & O \end{pmatrix}$$

For any $\sigma_\mathbf{A}$ in $S(V, Q)$ we compute

(g) $\mathbf{X}\sigma_\mathbf{A} = \mathbf{A}\mathbf{X}\mathbf{A}^{-1} \qquad \forall \mathbf{X} \in V$

This follows from $\mathbf{X}\sigma_\mathbf{A} = -\mathbf{X} + [g_Q(\mathbf{X}, \mathbf{A})/Q(\mathbf{A})]\mathbf{A} \overset{(f)}{=} -\mathbf{X} - g_Q(\mathbf{X}, \mathbf{A})\mathbf{A}^{-1} \overset{(c)}{=}$

$-\mathbf{X} + (\mathbf{X}\mathbf{A} + \mathbf{A}\mathbf{X})\mathbf{A}^{-1} = \mathbf{A}\mathbf{X}\mathbf{A}^{-1}$.

Let $S'(V, Q)$ be the subset of $S(V, Q)$ consisting of all those reflections $\sigma_\mathbf{A}$ for which

$$\mathbf{A} = \begin{pmatrix} O & E \\ B & O \end{pmatrix}$$

for any $B \neq O$ in K or

$$\mathbf{A} = \begin{pmatrix} -E & O \\ A & E \end{pmatrix}$$

for any A in K. Then we prove

(19) $S'(V, Q)$ *generates the group* $O(V, Q)$.

It suffices to prove that $S'(V, Q)$ generates $S(V, Q)$. Let $\sigma_{\mathbf{X}}$ be a reflection not in $S'(V, Q)$; then for

$$\mathbf{X} = \begin{pmatrix} A_1 & A_2 \\ A_3 & -A_1 \end{pmatrix}$$

we have $A_2 \neq O$ and $B := (A_3 A_2 + A_1{}^2) A_2{}^{-2} = -(\det \mathbf{X}) A_2{}^{-2} \neq O$. We put $A := A_1 A_2{}^{-1}$ and obtain for

$$\mathbf{A} := \begin{pmatrix} -E & O \\ A & E \end{pmatrix}$$

and

$$\mathbf{B} := \begin{pmatrix} O & E \\ B & O \end{pmatrix}$$

$\sigma_{\mathbf{X}} = \sigma_{\mathbf{A}} \sigma_{\mathbf{B}} \sigma_{\mathbf{A}}$.

Let $\bar{\alpha}$ for α in $O(V, Q)$ denote the restriction of the collineation induced by α to $L(V) \backslash L(V, Q)$ and let $\bar{N} := \{\bar{\alpha} \mid \alpha \in N\}$ for $N \subset O(V, Q)$. Then $\bar{N}$ is a set of bijective mappings of $L(V) \backslash L(V, Q)$ onto itself.

(20) *We have* $O(V, Q) \cong \overline{O(V, Q)}$.

We consider the map

$$\varepsilon : \begin{cases} O(V, Q) \to \overline{O(V, Q)} \\ \alpha \mapsto \bar{\alpha} \end{cases}$$

Obviously, ε is a homomorphism. Moreover, we show $\ker \varepsilon = \{1\}$. Let

$$\mathbf{A} := \begin{pmatrix} O & E \\ O & O \end{pmatrix}$$

$$\mathbf{B} := \begin{pmatrix} O & O \\ E & O \end{pmatrix}$$

$$\mathbf{C} := \begin{pmatrix} E & E \\ -E & -E \end{pmatrix}$$

$$\mathbf{D} := \begin{pmatrix} E & X^{-1} \\ -X & -E \end{pmatrix} \quad \text{with} \quad X \neq O, E$$

Then $\mathbf{A}$, $\mathbf{B}$, $\mathbf{C}$, and $\mathbf{D}$ are isotropic vectors. Any three of them are linearly independent, and therefore form a basis of V. If $\alpha \in \ker \varepsilon$, then there exist elements A, B, C, D in K such that $\mathbf{A}\alpha = A\mathbf{A}$, $\mathbf{B}\alpha = B\mathbf{B}$, $\mathbf{C}\alpha = C\mathbf{C}$, $\mathbf{D}\alpha = D\mathbf{D}$, since $\bar{\alpha} = 1$. From $\det \alpha = 1$ we see that for any basis in the set of the four vectors $\mathbf{A}$, $\mathbf{B}$, $\mathbf{C}$, $\mathbf{D}$, $ABC = E = ABD = BCD = ACD$ and therefore $A = B = C = D$. Furthermore, from $(\mathbf{A} + \mathbf{B})\alpha = A(\mathbf{A} + \mathbf{B})$ follows $A^2 = E$, since $\mathbf{A} + \mathbf{B}$ is nonisotropic, hence $A = B = C = A^2 C = ABC = E$ and therefore $\alpha = 1$. Thus ε is an isomorphism of $O(V, Q)$ onto $\overline{O(V, Q)}$.

(21) (G^*, S^*) *is isomorphic to* $(O(V, Q), S(V, Q))$.

We consider the map

$$\psi : \begin{cases} T \to L(V) \backslash L(V, Q) \\ X \mapsto \langle \mathbf{X} \rangle \end{cases} \text{ such that } \mathbf{X} := \begin{cases} \begin{pmatrix} E & X^{-1} \\ -X & -E \end{pmatrix} & \text{if } X \neq O, U \\ \begin{pmatrix} O & O \\ E & O \end{pmatrix} & \text{if } X = U \\ \begin{pmatrix} O & E \\ O & O \end{pmatrix} & \text{if } X = O \end{cases}$$

Obviously, ψ is a bijective map. We show for any s in S':

(h) $\quad \psi^{-1}\bar{s}^*\psi = \bar{\sigma}_{\mathbf{A}} \quad \text{with} \quad \mathbf{A} = \begin{cases} \begin{pmatrix} -E & O \\ O^s & E \end{pmatrix} & \textit{if } s \in U; \\ \begin{pmatrix} O & E \\ E^s & O \end{pmatrix} & \textit{if } s \in Z. \end{cases}$

From the definition of the addition and the multiplication in K we see for any s in S':

$$X^s = -X + O^s \qquad \forall s \in U; \quad \forall X \in K$$
$$X^s = X^{-1}E^s \qquad \forall s \in Z; \quad \forall X \neq O \quad \text{in } K$$

Hence an easy multiplication of matrices implies $(X\bar{s}^*)\psi = (X^s)\psi = (X\psi)\bar{\sigma}_{\mathbf{A}}$ for any end X in T, as required for (h).

By definition of S' and $S'(V, Q)$ we obtain from (h) and (17) and (19):

(i) $\quad \psi^{-1}\overline{(S')}^*\psi = \overline{S'(V, Q)} \quad and \quad \psi^{-1}\bar{G}^*\psi = \overline{O(V, Q)}$.

By (i) the groups $\bar{G}^*$ and $\overline{O(V, Q)}$ are isomorphic, hence by (18) and (20) the groups G^* and $O(V, Q)$ are isomorphic. The product ρ of the map δ

and the map

$$\bar\psi : \begin{cases} \bar{G}^* \to \overline{O(V,Q)} \\ \bar\alpha^* \mapsto \psi^{-1}\bar\alpha^*\psi \end{cases}$$

and the map ε^{-1} is an isomorphism of G^* and $O(V, Q)$, mapping S^* onto $S(V, Q)$.

Last, we prove the following statement about the center $Z(G)$ of an S-group (G, S) whose group plane is a hyperbolic-metric plane [compare (11)]:

(22) *If there exists a polar triangle, then* $Z(G) = \{1\}$; *if there exists a central line* z, *then* $Z(G) = \{1, z\}$.

The first statement follows immediately from (12) and Theorem 4.4(b). If there is a central line z, then (12) yields that $Z(G) = (S^2 \cap Z(G)) \cup ((zS^2) \cap Z(G)) = \{1\} \cup ((zS^2) \cap Z(G))$ by Theorem 4.4(b). Since we see from $\alpha \in (zS^2) \cap Z(G)$ that $z\alpha \in S^2 \cap Z(G)$, we have $z\alpha = 1$ and $\alpha = z$. Thus $Z(G)$ is the cyclic group of order 2.

Now we are able to formulate and prove the basic theorems on hyperbolic-metric planes.

THEOREM 6.25. *For an S-group plane* E *the following are equivalent up to isomorphisms:*

(a) E *is a hyperbolic-metric plane.*

(b) E *is a hyperbolic-metric coordinate plane.*

(c) E *is a pappian oval-complement.*

(d) E *is an oval-complement.*

PROOF. We prove the theorem by going along the "circle" (a) $\Rightarrow$ (b) and (b) $\Rightarrow$ (c) and (c) $\Rightarrow$ (d) and (d) $\Rightarrow$ (a).

(a) $\Rightarrow$ (b): Let $E(G, S)$ be a hyperbolic-metric plane and (V, Q) the metric vector space over a field of ends in $E(G, S)$, as constructed before. We recall that dim $V^{\perp} \leq 1$ and Ind $(V, Q) = 1$ and dim rad $(V, Q) = 0$. Thus $I(V, Q)$ is a hyperbolic-metric coordinate plane. Let ρ denote the isomorphism constructed to prove (21). By (22) the assumptions of Theorem 4.11 are fulfilled. Applying (22) and equation ($\circ$) after Theorems 4.11 and 3.6, we have the following chain of equivalent statements [μ denotes the bijective map of $S(V, Q)$ onto $S'(V, Q)$, defined in Chapter 3, Section 6]:

$$(a, b, c) \in \kappa(G, S) \Leftrightarrow abc \in S \Leftrightarrow a^*b^*c^* \in S^*$$
$$\Leftrightarrow (a^*\rho)(b^*\rho)(c^*\rho) \in S^*\rho = S(V, Q)$$
$$\Leftrightarrow (a^*\rho\mu)(b^*\rho\mu)(c^*\rho\mu) \in S(V, Q)\mu = S'(V, Q)$$
$$\Leftrightarrow (a^*\rho\mu\Phi(V, Q)^{-1}, b^*\rho\mu\Phi(V, Q)^{-1}, c^*\rho\mu\Phi(V, Q)^{-1}) \in \kappa(V, Q)$$

Hence the map $a \mapsto a^*\rho\mu\Phi(V, Q)^{-1}$ is an isomorphism of $E(G, S)$ onto $I(V, Q)$. Hence $E(G, S)$ is isomorphic to a hyperbolic-metric coordinate plane.

(b) $\Rightarrow$ (c): Let $I(V, Q)$ denote a hyperbolic-metric coordinate plane. Then $I(V, Q)$ is isomorphic to an S-group plane by Theorem 4.9. Moreover, it is a substructure of the pappian coordinate incidence structure $I(V)$, and by Proposition 2.4 an oval-complement. Thus up to isomorphisms $I(V, Q)$ is an S-group plane that is a pappian oval-complement.

(c) implies (d), trivially.

(d) $\Rightarrow$ (a): Let $E(G, S)$ denote an S-group plane that is an oval-complement. Then $E(G, S)$ is a complete S-group plane. By Theorems 6.13, 6.15, and 6.21 $E(G, S)$ is not an elliptic plane or a Euclidean plane or a Strubecker plane. Moreover, Theorem 6.28 proved independently from this section implies $E(G, S)$ is also not a Minkowskian plane. Hence $E(G, S)$ is a hyperbolic-metric plane.

This concludes the proof of Theorem 6.25. ∎

Clearly, there is a direct proof that (d) implies (a). However, we leave it as an exercise.

From (21) and (22) we obtain immediately:

THEOREM 6.26. *Let* (G, S) *be an S-group whose S-group plane is a hyperbolic-metric plane, and let* (V, Q) *denote the metric vector space over the field* K *of ends constructed above. Then* (G^*, S^*) *is isomorphic to* $(O(V, Q), S(V, Q))$, *and the center of the group* G *is equal to* $\{1\}$ *or the cyclic group of order 2.*

Theorems 6.25 and 6.26 contain a proof of the main theorems formulated in the beginning of this chapter.

By the well known fact that the orthogonal group $(O(V, Q), S(V, Q))$ of a metric vector space (V, Q) over a field K such that $\mathrm{Ind}(V, Q) = 1$ and $\dim \mathrm{rad}(V, Q) = 0$ and $\dim V^\perp \leq 1$ is isomorphic to the group

$(PGL_2(K), J)$, where J denotes the set of all involutions, we can add the following:

COROLLARY 6.27. *If (G, S) is an S-group whose group plane is a hyperbolic-metric plane, then (G^*, S^*) is isomorphic to the group $(PGL_2(K), J)$ for the field K of ends constructed above.*

8. THE MINKOWSKIAN PLANES

The last case of complete S-group planes, the Minkowskian planes, can be treated along a line which may be considered as a combination of our elaborations in Section 5 concerning the Euclidean planes and the "Endenrechnung" in hyperbolic-metric geometry given in Section 7. We separate our treatment into several steps.

Let $E(G, S)$ denote a Minkowskian plane. By Axiom EM there exist at least two distinct 2-Δ-connected points, and Axiom $\Delta 1$ yields that there is at least one 1-Δ-connected point. Thus we see from Proposition 6.8(c):

(1) *Any two distinct 2-Δ-connected points are not connected by a line.*

From Proposition 6.8(e) we deduce:

(2) *Given any line a there exists exactly one polar point A, and we have $A = a^{\perp}$ or $A = a^{\perp} \cup \{a\}$.*

We denote the uniquely determined polar point to the line a by S_a. Thus we have $S_a = a^{\perp}$ or $S_a = a^{\perp} \cup \{a\}$. Obviously, $a^x = a$ for all x in S_a. Moreover, Proposition 6.8(e) and (f) yield:

(3) *The set of 2-Δ-connected points coincides with the set of polar points.*

Now we denote by P the set of all points in the Minkowskian plane $E(G, S)$ which are not 2-Δ-connected. The set P contains at least one 1-Δ-connected point, by Axiom $\Delta 1$. Our first goal is to show that $(P, S, \in)$ may be considered a substructure of an affine plane. To do this we introduce a relation of parallelism of lines in S. As in the case of Euclidean planes we define

$$a \quad \text{parallel to} \quad b \Leftrightarrow a = b \quad \text{or} \quad S(ab) \quad \text{is 2-}\Delta\text{-connected} \qquad (\circ)$$

Then two lines a, b are parallel if and only if they are identical or they have no point in P in common. The relation of parallelism of lines a, b is denoted by $a \parallel b$. Obviously, this relation is an equivalence relation, and the equivalence classes, which we call, as usual, parallel pencils, are the 2-Δ-connected points. Hence we have

(4) *The set of 2-Δ-connected points coincides with the set of parallel pencils.*

As in Section 5 we can show

(5) *Given any point A in P and any line b, there is exactly one line a such that* $a \in A$ *and* $a \| b$.

We remark that the triplet $(P, S, \in)$ is not an affine plane, since Axiom A1 is not valid. However, Axioms A2 and A3 hold [compare (5) and the proof of A3 in the case of Euclidean planes].

From Propositions 5.11, 6.8(a), and 6.7(c) we deduce

(6) *Given any point A in P and any line a not in A, there is exactly one line b in* $A \cap a^{\perp}$.

(7) *Given any point A in P and any line b, there is exactly one line a such that* $a \in A \cap S_b$.

Moreover, we can show

(8) *For any point A in P and any line x we have*

$$A^x = A \Rightarrow x \in A$$

PROOF. If $A^x = A$ and $x \notin A$, then $A \subset x^{\perp}$, and therefore by (2) $A = S_x$, which contradicts (3). Hence $A^x = A$ implies $x \in A$.

Next we prove

(9) *For A, B in P we have*

$$A \cap B \neq \varnothing \Leftrightarrow B \in A^S$$

PROOF. (A) Let $B \in A^S$ and thus $B = A^s$ for a suitable s. If $A = B$, then $A \cap B \neq \varnothing$, trivially. If $A \neq B$, then by (7) there is a line a in $A \cap S_s$. Obviously, $a \in A$, A^s and therefore $A \cap B = A \cap A^s \neq \varnothing$.

(B) Let $A \cap B \neq \varnothing$ and $c \in A$, B. If $|A| > 2$, then there is a line a in A such that $ac \neq ca$, since $A \neq S_c$ by (2), and therefore at most one line x in A is in $c^{\perp}$. Let D denote the pencil of lines parallel to a, and b a line through B and parallel to a. Then S_c and D are 2-Δ-connected points by

(3) and (4), and are not connected by (1). Hence Lemma 6.5 yields an element α in S_c^2 such that $a^\alpha = b$. Let d be the line in $A \cap S_c$ [compare (7)] and put $s := d\alpha$. Then $A^s = A^{d\alpha} = A^\alpha = S(ac)^\alpha = S(a^\alpha c) = S(bc) = B$ and therefore $B \in A^S$. If $|A| = 2$, then $|S_c| = 3$ and for $a \in A \cap S_c$ and $b \in B \cap S_c$ we have $a^s = b$ for $s \neq a$, b in S_c. This yields $A^s = S(ac)^s = S(a^s c) = S(bc) = B$ and $B \in A^S$ (compare Chapter 8 to this part of the proof).

By (9) any two points in P are connected if and only if there is a reflection mapping one point onto the other. If the points in P are not connected, there is a rotation mapping one point onto the other:

(10) *For A, B in P we have*

$$A \cap B = \varnothing \Rightarrow B \in A^{S^2}$$

To prove this, let $A \cap B = \varnothing$ and a be any line in A. By Proposition 6.8(d) we can choose a line b in B such that $S(ab)$ is not 2-Δ-connected. Then we deduce from (9) that there are lines s, t such that $A^s = S(ab)$ and $S(ab)^t = B$. Hence $B = A^{st} \in A^{S^2}$.

Items (9) and (10) contain that, given any two points in P, there is a motion of $E(G, S)$ mapping the one point onto the other. Since P contains at least one 1-Δ-connected point, this implies

(11) *Any point in P is 1-Δ-connected.*

Furthermore, we show

(12) *For A, B, C in P we have:* $A \cap B = \varnothing = A \cap C$ *and* $B \cap C \neq \varnothing$ *and* $B \neq C$ *for A, B, C $\in$ P imply* $C \in B^A$.

To prove this let $a \in B \cap C$. Then we see from (7) that there is a line b in $A \cap S_a$. If $B^b = B$ and therefore $C^b = C$, then b would be a central line, in contradiction to Proposition 6.7(a). Thus $B^b \neq B$ and therefore $B^b = C$,

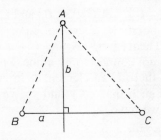

since $a \in B$, B^b, C and B, B^b, C are not connected with the Δ-connected point A. Hence $C \in B^A$.

(13)   *If* A, B, C *are points in* P *such that no two of them are connected by a line, then* $C \in B^{A^2}$.

To prove this, we choose any line a in A. Then there is a line b in $B \cap S_a$ by (7). We have $B^a \neq B$ by (7). If $B^a \cap C = \varnothing$, then (12) yields that there is a line c in C such that $B^a = B^c$. Clearly, $a \neq c$ and $S(ac) = S_b$, and there is a line $d \in B \cap S_b$ by (7). Thus we have $B^{dac} = B^{ac} = B$ and therefore $dac \in B$ by (8). From $d, dac \in B$, $S(ac)$ and $B \neq S(ac)$ (since

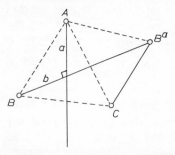

$a \notin B$) we see that $d = dac$ and thus $a = c \in A$, C, a contradiction. Hence $B^a \cap C \neq \varnothing$ and $B^a \neq C$ [since B, C are not connected and B, B^a are connected, by (9)], and therefore (12) yields $C \in B^{aA} = B^{A^2}$.

We recall that for any two Δ-connected points A, B not connected by a line the set $[A, B]$ is the set of all the points B^σ for a suitable mapping $\sigma := \sigma_{qBA}$. For these sets we prove

(14) $[A, B] = B^A$.

The proof contains two parts: (A) We have $B^A \subset [A, B]$. Let a be any line in A; then $a \notin B$, and the mapping $\sigma := \sigma_{aBA}$ is defined. Obviously, $x_A = a$ for all x in B, and therefore $x\sigma = axx_A = x^a$ for all x in B. Thus the point B^σ exists and is equal to B^a. Hence $B^a \in [A, B]$ and therefore $B^A \subset [A, B]$.

(B) We prove $[A, B] \subset B^A$. Let $C \in [A, B]$ and $C = B^\sigma$ and $\sigma = \sigma_{qBA}$. If we assume $C \notin B^A$, the points B, C are not connected by a line, by (12), or we have $B = C$. By Theorem 5.7 A and C are not connected, and by definition A and B are not connected by a line. From (13) or directly we see that $C \in B^{A^2}$. Thus there is an element α in A^2 such that $B^\alpha = C$.

We choose x, y in B such that $x \neq y$ and $x\sigma$, $y\sigma$ exist (this is possible, since B^σ exists). Then we have

(a) $(x\sigma)(y\sigma) = (xy)^{x_A}(x_A y_A)$.

Therefore $x_A y_A \neq 1$ [otherwise $C = S(x\sigma y\sigma) = S(xy)^{x_A} = B^{x_A} \in B^A$] and $A = S(x_A y_A)$. We have $x_A \alpha^{-1} \in A$, and by (9) B and $B^{x_A \alpha^{-1}}$ are connected by a line. Thus $C = B^\alpha$ and B^{x_A} are connected by a line, say z. Then $z(xy)^{x_A} \in S$ and $z(x\sigma)(y\sigma) \in S$. From (a) we see that $z(xy)^{x_A} \in A$, B^{x_A} and therefore $A \cap B^{x_A} \neq \varnothing$ and thus $A \cap B \neq \varnothing$, a contradiction. Thus our assumption $C \notin B^A$ is false, and (14) is proved.

(15) *Any two distinct points in a set* $[A, B]$ *for* $A \in P$ *are not connected.*

To prove this, let U, V be two distinct points in $[A, B]$. Then we deduce from (14) that there are lines a, c in A such that $U = B^a$ and $V = B^c$. If $U \cap V \neq \varnothing$, then we see from (12) there is a line b in A such that $U^b = V$. We have $B^{abc} = U^{bc} = V^c = B$, and therefore by (8) $abc \in B$. Thus $abc \in A$, B, in contradiction to the assumptions. Hence U, V are not connected.

Now we wish to introduce ideal lines. Let A be any point in P, and B a point not connected with A. Then $B \in P$ by Proposition 6.7(c). We put

$$\langle A, B \rangle := \{A\} \cup \{B\} \cup \{X \mid X \quad \text{is not connected with} \quad A \text{ and } B\}$$

Any set $\langle A, B \rangle$ of points is called an *ideal line* of the Minkowskian plane $E(G, S)$. Obviously, we have

(16) $\langle A, B \rangle = \langle B, A \rangle$.

Moreover, Proposition 6.7(c) implies

(17) $\langle A, B \rangle \subset P$.

Furthermore, we show

(18) $\langle A, B \rangle = \{A\} \cup B^{A^2} = \{A\} \cup [A, B^a]$ *for all* a *in* A.

To prove this, by (13) we have $\langle A, B \rangle \subset \{A\} \cup B^{A^2}$. From (14) we see that $\{A\} \cup B^{A^2} = \{A\} \cup B^{aA} = \{A\} \cup [A, B^a]$ for all a in A. By Theorem 5.7 any point $X \neq B$ in $[A, B^a]$ is not connected with A, and by (15) not connected with $(B^a)^a = B$, since $(B^a)^a \in [A, B^a]$ by (14). Thus $\{A\} \cup [A, B^a] \subset \langle A, B \rangle$. Hence we have

$$\langle A, B \rangle \subset \{A\} \cup B^{A^2} = \{A\} \cup [A, B^a] \subset \langle A, B \rangle$$

which implies (18).

From Theorem 5.7 and (18) and (15) we deduce immediately

(19) *Any two distinct points in* $\langle A, B \rangle$ *are not connected by a line in* S.

Furthermore, we show

(20) $C \in \langle A, B \rangle$ *and* $C \neq A$ *implies* $\langle A, B \rangle = \langle A, C \rangle$.

The proof is as follows: By (18), to C in $\langle A, B \rangle$ there is an α in A^2 such that $C = B^\alpha$, and we have $\langle A, B \rangle = \{A\} \cup B^{A^2} = \{A\} \cup B^{\alpha A^2} = \{A\} \cup C^{A^2} = \langle A, C \rangle$.

(21) $X, Y \in \langle A, B \rangle, \langle C, D \rangle$ *implies* $X = Y$ *or* $\langle A, B \rangle = \langle C, D \rangle$.

The proof is as follows: If $X \neq Y$, then $X \cap Y = \varnothing$ by (19). Moreover, $X \in P$ by (17). Without loss of generality, we may assume $X \neq A, C$. Then we deduce from (20) $\langle A, B \rangle = \langle A, X \rangle = \langle X, Y \rangle = \langle C, X \rangle = \langle C, D \rangle$.
    By (18), Lemma 5.8, and (19), or by a direct argument, we have

(22)   *For* $\langle A, B \rangle$ *and* c *there is exactly one point* C *such that* $c \in C \in$ $\langle A, B \rangle$.

Let α^* be any motion of the Minkowskian plane $E(G, S)$. Then α^* induces a mapping of the ideal lines if we put

$$\langle A, B \rangle \alpha^* := \langle A, B \rangle^\alpha = \{ X^\alpha \mid X \in \langle A, B \rangle \}$$

By definition of an ideal line we have

(23) $\langle A, B \rangle^\alpha = \langle A^\alpha, B^\alpha \rangle \qquad \forall \alpha \in G$.

Thus an ideal line is mapped onto an ideal line. Motion α^* induces a bijective map of the set of all the ideal lines onto itself. Moreover, we can prove

(24)   *If* $\langle A, B \rangle \neq \langle A, C \rangle$, *then* $\langle A, B \rangle^A = \langle A, C \rangle$.

By (20) $\langle A, B \rangle \neq \langle A, C \rangle$ yields $C \notin \langle A, B \rangle$ and therefore $C \notin B^{A^2}$. By (13) B and C cannot be not connected, and therefore we see from (12) that there is an element b in A such that $B^b = C$. This implies by (18) that

$$\langle A, C \rangle = \{A\} \cup C^{A^2} = \{A\} \cup B^{bA^2} = \{A\} \cup B^{aA^2} = \langle A, B^a \rangle$$
$$= \langle A^a, B^a \rangle = \langle A, B \rangle^a$$

for all a in A, as required.

(25)   *Given any two ideal lines* $\langle A, B \rangle, \langle C, D \rangle$, *there exists an element* α *in* G *such that* $\langle A, B \rangle^\alpha = \langle C, D \rangle$.

By (9) and (10), to A, C there is an element β in G such that $A^\beta = C$. Then $\langle A, B \rangle^\beta = \langle C, B^\beta \rangle$, and therefore $\langle A, B \rangle^\beta = \langle C, D \rangle$ or $\langle A, B \rangle^{\beta c} = \langle C, D \rangle$ for any c in C by (24). Thus we may put $\alpha := \beta$ or $\alpha := \beta c$ and obtain $\langle A, B \rangle^\alpha = \langle C, D \rangle$.

(26) *For any ideal line* $\langle A, B \rangle$ *we have* $\langle A, B \rangle^{S^3} = \langle A, B \rangle^S$.

Let $abc \in S^3$. By (22) and (20) we may assume $a \in A$. Then (22) implies there is a point C such that $b \in C \in \langle A, B \rangle^a$ and a point D such that $c \in D \in \langle A, B \rangle^{ab}$. Since C is Δ-connected, there are elements a', b' in A, C, respectively, such that $a'b'c \in S$. We put $d := a'b'c$ and obtain from (24):

$$\langle A, B \rangle^{abc} = (((\langle A, B \rangle^a)^b)^c = (((\langle A, B \rangle^{a'})^b)^c$$
$$= (((\langle A, B \rangle^{a'})^{b'})^c = \langle A, B \rangle^d$$

This yields immediately (26).

We denote the set of all ideal lines by S_ω, and deduce from (26) and (25):

(27) $S_\omega = \langle A, B \rangle^S \cup \langle A, B \rangle^{S^2}$ *for any ideal line* $\langle A, B \rangle$.

Two ideal lines $\langle A, B \rangle$ and $\langle C, D \rangle$ are said to be *parallel*, denoted by $\langle A, B \rangle \| \langle C, D \rangle$, if they are equal or if $\langle A, B \rangle \cap \langle C, D \rangle = \varnothing$. We note

$$\langle A, B \rangle \| \langle C, D \rangle \Leftrightarrow \langle A, B \rangle = \langle C, D \rangle \quad \text{or} \quad \langle A, B \rangle \cap \langle C, D \rangle = \varnothing \qquad (\circ\circ)$$

We wish to prove the following criterion for the nonparallelism of ideal lines:

(28) $\langle A, B \rangle \nparallel \langle C, D \rangle \Leftrightarrow \langle C, D \rangle \in \langle A, B \rangle^S$.

This is proved in two parts. (A) Let $\langle A, B \rangle^s = \langle C, D \rangle$. By (22) there is a point U such that $s \in U \in \langle A, B \rangle$, and we have $U \in \langle A, B \rangle$, $\langle C, D \rangle$ since $U^s = U$. Thus $\langle A, B \rangle \| \langle C, D \rangle$ would imply $\langle A, B \rangle = \langle C, D \rangle$ and therefore $X^s \in \langle A, B \rangle$ for all X in $\langle A, B \rangle$. From (19) and (9) we see that $X^s = X$, and therefore by (8) $s \in X$ for all X in $\langle A, B \rangle$, a contradiction to (19). Hence $\langle A, B \rangle \nparallel \langle C, D \rangle$.

(B) Let $\langle A, B \rangle \nparallel \langle C, D \rangle$; then there is a point U in $\langle A, B \rangle \cap \langle C, D \rangle$, and we have $\langle A, B \rangle \neq \langle C, D \rangle$. Without loss of generality we may assume that $U \neq A$, C. Then (20) and (24) yield for all s in U:

$$\langle A, B \rangle^s = \langle U, A \rangle^s = \langle U, C \rangle = \langle C, D \rangle$$

We add a criterion for the parallelism of ideal lines:

(29) $\langle A, B\rangle \| \langle C, D\rangle \Leftrightarrow \langle C, D\rangle \in \langle A, B\rangle^{S^2}$.

By (27) we have $\langle C, D\rangle \in \langle A, B\rangle^S$ or $\langle C, D\rangle \in \langle A, B\rangle^{S^2}$. The first is equivalent to $\langle A, B\rangle \nparallel \langle C, D\rangle$ by (28). Hence $\langle C, D\rangle \in \langle A, B\rangle^{S^2} \Leftrightarrow \langle A, B\rangle \| \langle C, D\rangle$.

(30) *Parallelism of ideal lines is an equivalence relation. There are exactly two equivalence classes.*

By (27) the set S_ω of all ideal lines is equal to $\langle A, B\rangle^S \cup \langle A, B\rangle^{S^2}$ for any ideal line $\langle A, B\rangle$. From (26) and (29) we see that any two ideal lines are parallel if and only if they belong to the same set of the two sets $\langle A, B\rangle^S$ or $\langle A, B\rangle^{S^2}$. Hence we have exactly two classes of parallel ideal lines. This proves (30).

Any equivalence class of the relation of parallelism of ideal lines is called a *parallel pencil of ideal lines*.

If we add the set S_ω of ideal lines to the set S of lines in the Minkowskian plane, we arrive at an affine plane. More precisely, for a in S, let

$$g(a) := \{X \mid a \in X\}$$

and let L be the set

$$L := \{g(a) \mid a \in S\} \cup S_\omega$$

Then we can prove

(31) $(P, L, \in)$ *is an affine plane.*

For A1: The existence of a line connecting two points is clear by definition. The uniqueness follows immediately from that in $E(G, S)$ and by (19) and (21).

For A2: By (5) we only have to discuss the case that the line is an ideal line. Let A be any point and $\langle B, C\rangle$ any ideal line. By (9) and (10) there is an element α in S or in S^2 such that $B^\alpha = A$. If $\alpha \in S$, then we choose any element a in A and obtain $\alpha a \in S^2$ and $B^{\alpha a} = A$. Thus we may assume in any case that $\alpha \in S^2$. Then $\langle B, C\rangle^\alpha \| \langle B, C\rangle$ and $A = B^\alpha \in \langle B, C\rangle^\alpha$. This proves the existence of a line parallel to $\langle B, C\rangle$ through A. Since no line $g(a)$ for $a \in S$ is parallel to any ideal line by (22), the uniqueness of the line parallel to $\langle B, C\rangle$ through A can be seen from: $A \in \langle B', C'\rangle, \langle B'', C''\rangle$ and $\langle B', C'\rangle, \langle B'', C''\rangle \| \langle B, C\rangle$ implies $\langle B', C'\rangle \| \langle B'', C''\rangle$ by (30), and therefore, by definition, $\langle B', C'\rangle = \langle B'', C''\rangle$.

A3 follows as in the Euclidean case.

The affine plane $(P, L, \in)$ is called the *affine extension* of the Minkowskian plane $E(G, S)$, and is denoted by $AI(G, S)$.

(32) *Any element α in the group G induces a collineation of the affine extension* AI(G, S).

Let α be any element in G; then the map

$$\alpha^\circ : \begin{cases} P \to P \\ A \mapsto A^\alpha \end{cases}$$

is bijective and maps any line in L onto a line by (23). Hence it is a collineation of $AI(G, S)$.

In the following, for any element α in the group G the collineation induced by it is denoted by α°.

(33) *If* a$\|$b, *then* (ab)$^\circ$ *is a translation of* AI(G, S).

If $a \| b$, then $a = b$ or $S(ab)$ is 2-Δ-connected. Hence Lemma 6.5 yields $A^{ab} = A$ for any 2-Δ-connected point A. Thus by (4) any parallel pencil of lines in S is mapped onto itself by $(ab)^\circ$. Moreover, (29) implies that any parallel pencil of ideal lines in S_ω is mapped onto itself. Hence $(ab)^\circ$ is a homothety (compare the Appendix). If $a \neq b$, then $(ab)^\circ$ has no fixed point: Let $A^{ab} = A$ and a' be the line parallel to a through A and $b' := a'ab$. Then $A^{b'} = A^{a'b'} = A^{ab} = A$, and therefore by (8) $b' \in A$ and $A = S(a'b') = S(ab)$, which contradicts the statements $A \in P$ and $S(ab)$ is 2-Δ-connected. Thus $(ab)^\circ$ is a translation.

(34) *The translation group of the affine extension* AI(G, S) *is linearly transitive.*

Let A, B denote two distinct points. If $A \cap B \neq \varnothing$ and $v \in A, B$, then we see from (9) there is a line b such that $A^b = B$. Obviously, $b \in v^\perp$, and by (7) there is a line $a \in A \cap S_v$. Then $A^{ab} = B$ and $a, b \in S_v$. From (3) and (4) we deduce $a \| b$, and therefore $(ab)^\circ$ is a translation mapping A onto B.

If $A \cap B = \varnothing$, then there are lines s, t in A, B, respectively, such that $S(st) \in P$. Then there is a translation τ mapping A onto $S(st)$ and a translation τ' mapping $S(st)$ onto B, by our preceding discussion. Hence $\tau\tau'$ is a translation mapping A onto B.

Now we wish to establish an analogue to the "Endenrechnung" in hyperbolic-metric geometry and to the introduction of a coordinate field for a Desarguesian affine plane (compare the Appendix).

By Proposition 6.7(c) and (11) there exist in P at least two points, say O and E, not connected by a line. We put

$$K := \langle O, E \rangle$$

To define an addition in K and a multiplication in K, we consider the group T of translations with fixed lines parallel to $\langle O, E \rangle$ and the group

$$D := \{ \alpha^\circ \mid \alpha \in O^2 \}$$

of collineations of $AI(G, S)$. Obviously, every translation in T maps $K = \langle O, E \rangle$ onto itself. The same is true for any collineation α° in D, since $O \in \langle O, E \rangle$, $\langle O, E \rangle^\alpha$ and $\langle O, E \rangle \| \langle O, E \rangle^\alpha$ by (29).

From (34) and the Appendix we see that T is linearly transitive and abelian. Thus for any A in K there is exactly one translation τ in T mapping O onto A. We denote this translation by τ_A.

The group D is abelian and, similarly to T, we can show

(35)   *To any* $A \neq O$ *in* K *there exists exactly one collineation* α° *in* D *such that* $E\alpha^\circ = A$.

The existence of α° follows from (29) and (13). To prove the uniqueness, we show: $E^\alpha = E^\beta$ and $\alpha, \beta \in O^2$ implies $E^{\alpha\beta^{-1}} = E$ and $\alpha\beta^{-1} \in O^2$, and therefore $E^a = E^b$ for $\alpha\beta^{-1} = ab$ and $a, b \in O$. Since $a, b \notin E$, we have $E \neq E^a = E^b$. We deduce from (6) there is a line s in $E \cap a^\perp$. Obviously, $a, b \in s^\perp$ and therefore $a = b$, since $O \neq S_s$ by (3). Thus $\alpha\beta^{-1} = 1$ and $\alpha = \beta$.

We denote the uniquely determined element α° in D such that $E\alpha^\circ = A$ by α_A. We have

(36)   *If* $\alpha \in O^2$ *and* τ_A *in* T, *then* $(\alpha^\circ)^{-1}\tau_A\alpha^\circ = \tau_{A^\alpha}$.

Since α° is a collineation, $(\alpha^\circ)^{-1}\tau_A\alpha^\circ$ is a translation. From $O(\alpha^\circ)^{-1}\tau_A\alpha^\circ = O\tau_A\alpha^\circ = A^\alpha$ we see that $(\alpha^\circ)^{-1}\tau_A\alpha^\circ = \tau_{A^\alpha}$.

We define

$$\begin{cases} A + B := O\tau_A\tau_B & \text{for } A, B \in K \\ A \cdot B := E\alpha_A\alpha_B & \text{for } A, B \neq O \text{ in } K \\ A \cdot O := O \cdot A := O & \text{for } A \text{ in } K \end{cases} \qquad (\circ\circ\circ)$$

Obviously, we have

$$\tau_{A+B} = \tau_A\tau_B \qquad \forall A, B \in K$$

$$\alpha_{A \cdot B} = \alpha_A\alpha_B \qquad \forall A, B \neq O \text{ in } K$$

Hence $(K, +)$ is a group isomorphic to T, and $(K \setminus \{O\}, \cdot)$ is a group isomorphic to D. We prove one of the distributive laws as follows:

If $C = O$, then $(A + B) \cdot C = O = O + O = A \cdot C + B \cdot C$. If $C \neq O$, then we compute

$$
\begin{aligned}
(A + B) \cdot C &= (A + B)\alpha_C = O\tau_{A+B}\alpha_C = O\alpha_C^{-1}\tau_{A+B}\alpha_C \\
&= O\alpha_C^{-1}\tau_A\tau_B\alpha_C = O\alpha_C^{-1}\tau_A\alpha_C\alpha_C^{-1}\tau_B\alpha_C \\
&= O\tau_{A\alpha_C}\tau_{B\alpha_C} = O\tau_{A \cdot C}\tau_{B \cdot C} = O\tau_{A \cdot C + B \cdot C} \\
&= A \cdot C + B \cdot C
\end{aligned}
$$

Thus we have proved

(37) $(K, +, \cdot)$ is a field.

We call this field a *coordinate field of the Minkowskian plane.*

Now we wish to show that the affine extension $AI(G, S)$ is isomorphic to the affine coordinate plane $A(K)$ over the coordinate field of the Minkowskian plane. We recall (compare the Appendix) that the points of $A(K)$ are the pairs $[A, B]$ of elements in K, thus the vectors in the two-dimensional vector space $K^2 := K \times K$ over K. The lines of $A(K)$ are the residue classes $[A, B] + \langle [C, D] \rangle$ of the one-dimensional subspaces of K^2 (thus we have $[C, D] \neq [O, O]$).

To introduce coordinates in the affine extension $AI(G, S)$ of the Minkowskian plane, we have to choose two axes of a coordinate system. One of these axes is $K = \langle O, E \rangle$, and the other the uniquely determined ideal line K' distinct from K through O. By (24) we obtain $K^x = K'$ for all x in O.

Let A be any point in P. Then A is incident with exactly two ideal lines, one of them parallel to K, the other parallel to K'. These intersect

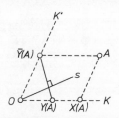

K' and K, respectively, in exactly two points, denoted by $\bar{Y}(A)$ and $X(A)$. We choose a fixed line s in O and put $Y(A) := (\bar{Y}(A))^s$. Then $X(A), Y(A) \in K$, and the map

$$
\Phi : \begin{cases} P \to K^2 \\ A \mapsto [X(A), Y(A)] \end{cases}
$$

is bijective. Here $X(A)$ is called the *x-coordinate*, and $Y(A)$ is called the *y-coordinate* of the point A. We wish to show

(38) *For all* A *in* K *and all* B *in* P *we have*

$$(B\tau_A)\Phi = B\Phi + A\Phi$$

Since τ_A maps any line parallel to $\langle O, E\rangle = K$ onto itself, we have $\bar{Y}(B\tau_A) = \bar{Y}(B)$ and therefore $Y(B\tau_A) = Y(B)$. Moreover, we have $X(B\tau_A) = X(B)\tau_A = X(B) + A$.   Hence $(B\tau_A)\Phi = [X(B\tau_A), Y(B\tau_A)] = [X(B) + A, Y(B)] = [X(B), Y(B)] + [A, O] = B\Phi + A\Phi$.

Now we can prove

(39) Φ *is an isomorphism of the affine extension* AI(G, S) *onto the affine coordinate plane* A(K) *over the coordinate field of the Minkowskian plane.*

We have to show that Φ maps the set of the points on a line in L onto a line in $A(K)$ [for $AI(G, S) = (P, L, \in)$].

Given any line g in L, we discuss successively several cases:

(1) g is an ideal line parallel to $\langle O, E\rangle$: Then g has exactly one point D in common with K', and we have $g = \{A \mid \bar{Y}(A) = D\}$ and therefore

   (b) $g\Phi = \{A\Phi \mid Y(A) = D^s\} = [O, D^s] + \langle[E, O]\rangle$.

(2) g is an ideal line parallel to K': Then g has exactly one point D in common with K and we have

   (c) $g\Phi = \{A\Phi \mid X(A) = D\} = [D, O] + \langle[O, E]\rangle$.

(3) g is a line $g(a)$ with $a \in O$: We put $\alpha := as$ and $B := E^\alpha$. Then $\alpha^\circ = \alpha_B$ and we obtain the following equivalent statements:

$$A \in g(a) \Leftrightarrow \bar{Y}(A)^a = X(A) \Leftrightarrow Y(A) = X(A)^\alpha$$
$$\Leftrightarrow Y(A) = X(A)\alpha_B = X(A) \cdot B$$
$$\Leftrightarrow A\Phi = X(A) \cdot [E, B]$$
$$\Leftrightarrow A\Phi \in \langle[E, B]\rangle.$$

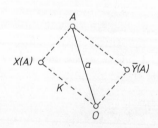

Hence we obtain

(d) $g\Phi = \langle[E, B]\rangle$.

(4) g is a line $g(a)$ with $a \notin O$. Then there is a translation τ in T such that $O \in g(a)\tau$. Since $g(a)\tau \| g(a)$, there is a line b in O such that $g(a)\tau = g(b)$. We put $C := O\tau$; thus $\tau = \tau_C$, and $B := E^{bs}$, to yield by (38) the following equivalent statements:

$$A \in g(a) \Leftrightarrow A\tau \in g(a)\tau \Leftrightarrow A\tau_C \in g(b) \Leftrightarrow (A\tau_C)\Phi \in \langle[E, B]\rangle$$
$$\Leftrightarrow A\Phi + C\Phi \in \langle[E, B]\rangle \Leftrightarrow A\Phi \in [-C, O] + \langle[E, B]\rangle.$$

This yields

(e) $g\Phi = [-C, O] + \langle[E, B]\rangle$.

Hence Φ induces an injective map of the set L of lines in the affine extension $AI(G, S)$ into the set of lines of the affine coordinate plane $A(K)$ over K. However, (b) to (d) show that any line in $A(K)$ is an image of a line in $AI(G, S)$ since every line in $A(K)$ is of the type

$$[O, A] + \langle[E, O]\rangle \quad \text{or} \quad [A, O] + \langle[O, E]\rangle \quad \text{or} \quad [A, O] + \langle[E, B]\rangle$$

with arbitrary A, B in K. Thus Φ is surjective, and (39) is proved. We conclude this chapter with the following theorem:

THEOREM 6.28. *For an S-group plane* E *the following are equivalent up to isomorphisms:*

(a)   E *is a Minkowskian plane.*

(b)   E *is a double star-complement.*

(c)   E *is a pappian double star-complement.*

(d)   E *is a Minkowskian coordinate plane.*

PROOF.   If $E = E(G_0, S_0)$, then our assertion holds trivially. Now let $E \neq E(G_0, S_0)$.

$(a) \Rightarrow (b)$. Let $E(G, S)$ denote a Minkowskian plane and $AI(G, S)$ its affine extension. Let (L', κ') be the projective incidence structure corresponding to the projective extension of $AI(G, S)$ (compare the Appendix). We may interpret $E(G, S)$ as a substructure of (L', κ'). Obviously, $L' \setminus S$ consists of the two parallel pencils of ideal lines and the line at infinity of $AI(G, S)$. Hence $E(G, S)$ is a double star-complement in (L', κ').

$(b) \Rightarrow (c)$. Let $E(G, S)$ be an S-group plane that is a double star-complement. Then by our remarks in Chapter 2, Section 3, there are 1-Δ-connected points and at least two 2-Δ-connected points which are not connected by a line, since the projective extension is not the minimal model of a projective incidence structure. Hence $E(G, S)$ is a Minkowskian plane, and its affine extension $AI(G, S)$ is a pappian affine plane, hence the projective extension of $AI(G, S)$ is a pappian projective plane (compare the Appendix). Thus $E(G, S)$ is a pappian double star-complement.

(c) $\Rightarrow$ (d). Let $E(G, S)$ denote an S-group plane that is a pappian double star-complement. As remarked above, $E(G, S)$ is a Minkowskian plane, and up to isomorphisms we may interpret $E(G, S)$ as a double star-complement in $I(V)$ over the coordinate field K of the Minkowskian plane $E(G, S)$. Let I_1 and I_2 denote the two "stars," containing the two parallel pencils of ideal lines of the Minkowskian plane $E(G, S)$.

We choose a basis $\mathbf{a}, \mathbf{b}, \mathbf{c}$ of V such that $\langle \mathbf{a} \rangle$ and $\langle \mathbf{b} \rangle$ are the ideal lines through the point O introduced to constitute the field K, and such that $\langle \mathbf{c} \rangle$ is the line at infinity of the affine extension $AI(G, S)$. We consider the map

$$Q : \begin{cases} V \to K \\ a\mathbf{a} + b\mathbf{b} + c\mathbf{c} \mapsto ab \end{cases} \quad (a, b, c \in K)$$

Then Q is a quadratic form and (V, Q) a metric vector space such that $\dim V^{\perp} = 1$ and $\mathrm{Ind}\,(V, Q) = 1$ and $\dim \mathrm{rad}\,(V, Q) = 1$ [we have $V^{\perp} = \mathrm{rad}\,(V, Q) = \langle \mathbf{c} \rangle$]. Thus $I(V, Q)$ is a Minkowskian coordinate plane.

Since $I_1 = \langle \mathbf{a}, \mathbf{c} \rangle$ and $I_2 = \langle \mathbf{b}, \mathbf{c} \rangle$, we have $L(V, Q) = S$, hence $E(G, S) = (S, \kappa(G, S)) = (L(V, Q), \kappa(V, Q)) = I(V, Q)$ and $E(G, S)$ is isomorphic to the Minkowskian coordinate plane $I(V, Q)$.

(d) $\Rightarrow$ (a). Let $I(V, Q)$ denote a Minkowskian coordinate plane. Then we deduce from Proposition 3.5 and Theorem 4.9 that $I(V, Q)$ is isomorphic to an S-group plane $E(G, S)$. Since $I(V, Q)$ is a double star-complement by Proposition 2.4, the S-group plane $E(G, S)$ is a double star-complement and therefore a Minkowskian plane by our preceding discussion. Thus $I(V, Q)$ is isomorphic to a Minkowskian plane. ∎

THEOREM 6.29. *Let* (G, S) *be an S-group whose group plane is a Minkowskian plane and let* (V, Q) *denote the metric vector space constructed to* $E(G, S)$ *in Theorem 6.28. Then* $(G^*, S^*) \cong (G, S)$ *and* (G, S) *is isomorphic to* $(O(V, Q), S(V, Q))$. *In particular, the center* $Z(G)$ *of the group* G *is equal to* 1.

PROOF. Suppose the assumptions of Theorem 6.29 are fulfilled. We first prove the last statement of our theorem: By Proposition 6.7(a) and Corollary 4.7 we have $Z(G) \cap (S \cup S^2 \cup S^3) = \{1\}$. If $Z(G) \cap S^4 \neq \{1\}$, then Proposition 6.12 implies that $E(G, S)$ is a Strubecker plane, hence by

Theorem 6.21 a star-complement, and we arrive at a contradiction to Theorem 6.28. Hence $Z(G) \cap S^4 = \{1\}$ and therefore, by the Reduction Theorem 5.1, $Z(G) = \{1\}$.*

From the last statement we deduce that $(G, S) \cong (G^*, S^*)$.

From our assumptions we may interpret $E(G, S)$ as the substructure $I(V, Q)$ of the projective coordinate incidence structure $I(V)$ over the coordinate field K of the Minkowskian plane. If α denotes any element in G, then α° is a collineation of the affine extension $AI(G, S)$ of $E(G, S)$ [compare (32)]. This collineation can be embedded in a collineation of $I(V)$ (as discussed in the Appendix) which is denoted by the same symbol α° as the affine collineation. If $\alpha = x \in S$, then x° is a collineation of $AI(G, S)$ with the axis $g(x)$. Moreover, x° is an involutory axial collineation, since $Z(G) = \{1\}$. Thus the projective continuation of x° is a projective reflection of $I(V)$ in the line x. Obviously, x° is a projective reflection in the point S_x. In the notations of the proof of Theorem 6.28 x° interchanges the points I_1 and I_2 in $I(V)$ [compare (28)].

Moreover, the projective reflection x° is an element of $S'(V, Q)$: Let $\langle x \rangle := x$; then $\mathbf{x}$ is nonisotropic, and by Lemma 3.4 we may identify the collineation $\bar{\sigma}_\mathbf{x}$ in $S'(V, Q)$ with the projective reflection in the line $\langle \mathbf{x} \rangle$ (compare Chapter 3, Section 6). This projective reflection interchanges the points I_1 and I_2: These points are the only two two-dimensional isotropic subspaces, and therefore $\bar{\sigma}_\mathbf{x}$ fixes I_1 and I_2 or interchanges these two points. However, $I_1 \bar{\sigma}_\mathbf{x} = I_1$ and $I_2 \bar{\sigma}_\mathbf{x} = I_2$ would imply $\bar{\sigma}_\mathbf{x} = 1$, since $g(x) \notin I_1, I_2$, and any perspective collineation fixing two distinct points not on the axis is the identity (compare the Appendix). From $\bar{\sigma}_\mathbf{x} = 1$ we deduce $\mathbf{x} \in V^\perp = \mathrm{rad}\,(V, Q)$ and $Q(\mathbf{x}) = 0$, a contradiction. Thus we have $I_1 \bar{\sigma}_\mathbf{x} = I_2$ and therefore $(I_1 \bar{\sigma}_\mathbf{x})x^\circ = I_1$ and $(I_2 \bar{\sigma}_\mathbf{x})x^\circ = I_2$. Since $(\bar{\sigma}_\mathbf{x})x^\circ$ is a perspective collineation with axis x, we obtain $\bar{\sigma}_\mathbf{x} x^\circ = 1$ and therefore $x^\circ = \bar{\sigma}_\mathbf{x} \in S'(V, Q)$.

Conversely, any projective reflection $\bar{\sigma}_\mathbf{x}$ in $S'(V, Q)$ coincides with a projective reflection x° with $x \in S$: We put $x := \langle \mathbf{x} \rangle$. Then x is not isotropic and therefore $x \in S$. As just proved, we have $\bar{\sigma}_\mathbf{x} = x^\circ$. Thus $S'(V, Q)$ is identical with the set S° of all the projective reflections x° such that $x \in S$. The map

$$\chi : \begin{cases} G \to G^\circ \\ \alpha \mapsto \alpha^\circ \end{cases}$$

is surjective, clearly. Since $Z(G) = \{1\}$, the map $\alpha \mapsto \alpha^*$ is injective, hence χ is injective and therefore bijective. Moreover, it is an isomorphism. Thus (G, S) is isomorphic to (G°, S°). As we just proved, S° coincides

---

* A direct proof of the statement $Z(G) = \{1\}$ is left as an exercise.

with $S'(V, Q)$ and therefore $G°$ with $O'(V, Q)$. By Chapter 3, Section 6, $(O'(V, Q), S'(V, Q))$ is isomorphic to $(O(V, Q), S(V, Q))$, and therefore (G, S) is isomorphic to $(O(V, Q), S(V, Q))$. Since $(G, S) \cong (G^*, S^*)$, the theorem is proved. ■

COROLLARY 6.30. *Let* $E(G, S)$ *be a Minkowskian plane and let* $I(V, Q)$ *be the Minkowskian coordinate plane belonging to* $E(G, S)$. *Then* $(S, \kappa(G, S), \Phi(G, S))$ *and* $(L(V, Q), \kappa(V, Q), \Phi(V, Q))$ *are isomorphic complete metric planes.*

PROOF. By Theorem 4.11 and Proposition 3.5 $(S, \kappa(G, S), \Phi(G, S))$ and $(L(V, Q), \kappa(V, Q))$ are complete metric planes. We continue the proof of Theorem 6.29. Identifying $E(G, S)$ and $I(V, Q)$, we saw that $S° := \{x° \mid x \in S\}$ and $S'(V, Q)$ are identical (compare the definition of $\alpha°$ for α in G on p. 123).

Thus for any $\langle \mathbf{x} \rangle$ in S we have* $\langle \mathbf{x} \rangle \Phi(G, S) = \langle \mathbf{x} \rangle^* = \langle \mathbf{x} \rangle° \mid_S = \bar{\sigma}_{\mathbf{x}} = \langle \mathbf{x} \rangle \Phi(V, Q)$ and therefore $\Phi(G, S) = \Phi(V, Q)$. Hence $(S, \kappa(G, S), \Phi(G, S)) = (L(V, Q), \kappa(V, Q), \Phi(V, Q))$, and Corollary 6.30 is proved. ■

9. MAIN THEOREM ON COMPLETE METRIC PLANES

We recall that a complete metric plane is a triplet (L, κ, Φ) such that (L, κ) is a Δ-connected incidence structure that contains a quadrilateral and for which Φ is a map of L into the set of axial collineations of (L, κ) satisfying the condition that $a\Phi$ is a collineation with axis a for all a in L and that for $\sigma_a := a\Phi$ and $S := \text{Im } \Phi$ condition [S] is valid. We repeat this condition:

$$\sigma_a \sigma_b \sigma_c \in S \Leftrightarrow (a, b, c) \in \kappa \qquad [\text{S}]$$

The collineation σ_a is called a *reflection in the line a*.

Thus a complete metric plane is an incidence structure whose points are all Δ-connected and which contains four lines no three of them concurrent and for which, given any line, there is a reflection in that line, such that the theorem of the three reflections holds.

The following main theorem characterizes entirely the complete metric planes:

MAIN THEOREM 6.31 (on complete metric planes). *Up to isomorphisms the complete metric planes are all the metric planes over three-dimensional metric vector spaces* (V, Q) *such that dim* $V^\perp \leq 1$.

---

* We have denoted by $\bar{\sigma}_{\mathbf{x}}$ as well the collineation of $I(V)$ induced by $\sigma_{\mathbf{x}}$ as the restriction of this collineation to $L(V, Q)$.

PROOF. (A) Let $M(V, Q)$ be the metric plane over any three-dimensional vector space (V, Q) such that dim $V^\perp \leq 1$. Then Proposition 3.5 implies that $M(V, Q)$ is a complete metric plane.

(B) Let (L, κ, Φ) be a complete metric plane and $S := \operatorname{Im} \Phi$. By Theorem 4.9, there is an S-group $(\bar{B}, \bar{S})$ associated to the complete metric plane (L, κ, Φ) (compare Chapter 4, Section 7). Then (L, κ) is isomorphic to the S-group plane $E(\bar{B}, \bar{S})$, and (L, κ) is Δ-connected and contains at least four lines, no three of them concurrent. This being true as well for $E(\bar{B}, \bar{S})$, we see that $E(\bar{B}, \bar{S})$ is a complete S-group plane. If $E(\bar{B}, \bar{S})$ is not a Strubecker plane, then $(\bar{S}, \kappa(\bar{B}, \bar{S}), \Phi(\bar{B}, \bar{S}))$ is isomorphic to the metric plane $(L(V, Q), \kappa(V, Q), \Phi(V, Q))$ over a three-dimensional metric vector space (V, Q): In the Minkowskian case this follows from Theorem 6.28 and Corollary 6.30. For the other cases we see from Theorems 6.13, 6.25, and 6.28 that $(\bar{S}, \kappa(\bar{B}, \bar{S}))$ is isomorphic to $(L(V, Q), \kappa(V, Q))$ for a suitable metric vector space. Identifying $(\bar{S}, \kappa(\bar{B}, \bar{S}))$ with $(L(V, Q), \kappa(V, Q))$, we deduce in the Euclidean and in the elliptic case from Theorem 5.14 that $\Phi(\bar{B}, \bar{S}) = \Phi(V, Q)$, and our assertion is true. If $(\bar{S}, \kappa(\bar{B}, \bar{S}))$ is a hyperbolic-metric plane, then we see, continuing the argument for (28), (20), and (21) in Section 7, that $x(a^* \delta) = (x\delta)\sigma_{a\delta}$ for all x in $L(V, Q)$ if δ denotes the isomorphism $y \mapsto y^* \rho \mu \Phi(V, Q)^{-1}$ considered in the proof of Theorem 6.25. Hence δ is an isomorphism of the complete metric plane $(\bar{S}, \kappa(\bar{B}, \bar{S}), \Phi(\bar{B}, \bar{S}))$ onto $(L(V, Q), \kappa(V, Q), \Phi(V, Q))$. Obviously, we have dim $V^\perp \leq 1$ for these metric vector spaces. By Corollary 4.10 $(\bar{S}, \kappa(\bar{B}, \bar{S}), \Phi(\bar{B}, \bar{S}))$ is isomorphic to (L, κ, Φ), hence we have

$$(L, \kappa, \Phi) \cong (L(V, Q), \kappa(V, Q), \Phi(V, Q))$$

as required.

The last case, assuming $E(\bar{B}, \bar{S})$ is a Strubecker plane, cannot occur, as otherwise $E(\bar{B}, \bar{S})$ would contain a dual parallelogram a, b, c, d [compare (4) in the proof of Proposition 6.12]. Then a, b, c are not collinear, and $abcd \in Z(G)$ and $a^* b^* c^* d^* = (abcd)^* = 1$ [by (3) in the proof of Proposition 6.12]. Since Corollary 4.10 yields $(\bar{S}, \kappa(\bar{B}, \bar{S}), \Phi(\bar{B}, \bar{S}))$, and (L, κ, Φ) are isomorphic, we would deduce from

$$a^* b^* c^* = (a\Phi(\bar{B}, \bar{S}))(b\Phi(\bar{B}, \bar{S}))(c\Phi(\bar{B}, \bar{S})) \in \bar{S}^* = \operatorname{Im} \Phi(\bar{B}, \bar{S})$$

that a, b, c are collinear, applying axiom [S]. Thus we would arrive at a contradiction.

This concludes the proof of Main Theorem 6.31. ∎

Now we wish to call special attention to some outstanding cases of Main Theorem 6.31 on complete metric planes.

We call a triplet (L, κ, Φ) consisting of a projective incidence structure (L, κ) and a mapping Φ of L into the set of axial collineations of (L, κ) such that $a\Phi$ is a collineation with axis a satisfying [S] a *projective S-plane*, and a triplet (L, κ, Φ) consisting of an affine incidence structure (L, κ) and a similar mapping Φ of L into the set of axial collineations of (L, κ) satisfying [S] an *affine S-plane*. Obviously, all projective S-planes and all affine S-planes are complete metric planes.

Let $E(G, S) = (S, \kappa(G, S))$ denote an elliptic plane. Then Theorem 4.11 yields that $(S, \kappa(G, S), \Phi(G, S))$ is an S-plane. Moreover, since any point of $E(G, S)$ is completely connected (by Proposition 6.10), $(S, \kappa(G, S), \Phi(G, S))$ is a complete metric plane. We call it as well an elliptic plane. Theorem 6.13 then implies:

THEOREM 6.32. *Up to isomorphisms the elliptic planes are all the projective S-planes.*

Analogously, if $E(G, S)$ is a Euclidean plane, the triplet $(S, \kappa(G, S), \Phi(G, S))$ is a complete metric plane by Theorem 4.11, which is called as well a Euclidean plane. Then we deduce immediately from Theorem 6.15:

THEOREM 6.33. *Up to isomorphisms the Euclidean planes are all the affine S-planes.*

The last two theorems contain remarkable results: Given any projective or affine plane such that for any line a there is a collineation σ_a with the axis a satisfying the theorem of the three reflections (Axiom [S]), then the plane is a pappian plane and, moreover, it is an elliptic or a Euclidean plane, respectively. Analogous theorems can be deduced from Theorems 6.25 and 6.28. The exact formulation is omitted here and is left to the reader.

10. POLAR POINTS IN COMPLETE S-GROUP PLANES

We wish to add some remarks on polar points in complete S-group planes. We emphasize that the use of special properties of polar points (such as Proposition 6.37) was not essential in our proofs of the main theorems on complete S-group planes. Only in Minkowskian geometry, in the proof of (9) in Section 8, did we apply the existence of a polar point to a line.

We prepare with a few small statements on polar points in an arbitrary S-group plane.

LEMMA 6.34. *Let* a *be any line in an arbitrary S-group plane such that* $a^\perp \cup \{a\}$ *is contained in a point. Then* $xy = yx$ *for all* x, y *in* $a^\perp \cup \{a\}$.

PROOF. Obviously, we have

$$a^\perp \cup \{a\} = \{x \mid ax = xa\}$$

Now let $a^\perp \cup \{a\}$ be contained in a point and $x, y \in a^\perp \cup \{a\}$. Then $ax = xa$ and $ay = ya$ and $xya = ayx$, hence $xy = yx$. ∎

LEMMA 6.35. *Let* z *be a central line in an S-group plane containing a quadrilateral. Then for any line* $a \neq z$ *we have*

$$S(az) = a^\perp \cup \{a\}$$

PROOF. By our assumption on the S-group plane, we have, applying Theorem 4.4(b), $z \notin S^2$. Then we can deduce the following equivalent statements:

$$x \in a^\perp \cup \{a\} \Leftrightarrow xa = ax \Leftrightarrow xa = ax; xz = zx; az = za$$

$$\Leftrightarrow xaz = zax \Leftrightarrow (\text{since } xaz \neq 1)zax \in J \Leftrightarrow x \in S(az)$$

Thus $a^\perp \cup \{a\} = S(az)$. ∎

LEMMA 6.36. *If* a, b, c *is a polar triangle in an arbitrary S-group plane, then*

$$a^\perp = S(bc)$$

PROOF. Since $abc = 1$, we have $a^\perp = \{x \mid ax \in J\} = \{x \mid bcx \in J\} = S(bc)$. ∎

Now let $E(G, S)$ denote any complete S-group plane. By Proposition 6.12, in a Strubecker plane there are no two lines a, b such that $a \perp b$. Thus for any line a in a Strubecker plane there does not exist a polar point; moreover, $a^\perp = \varnothing$. Thus we may restrict ourselves to the other cases of complete S-group planes. First, we prove:

PROPOSITION 6.37. *For any line* a *in a complete S-group plane which is not a Strubecker plane, the set* $a^\perp$ *or the set* $a^\perp \cup \{a\}$ *is a point, provided* a *is not a central line.*

PROOF. Obviously, our assertion holds for the two exceptional complete S-group planes $E(G_0, S_0)$ and $E(G_1, S_1)$.

Now let $E(G, S)$ be any complete S-group plane $\neq E(G_0, S_0)$, $E(G_1, S_1)$.

If $E(G, S)$ is a Euclidean or a Minkowskian plane, then our assertion follows immediately from Proposition 6.8(e).

If $E(G, S)$ is a hyperbolic-metric plane, then (11) in Section 7 yields the existence of a central line z or a polar triangle. In the first case the statement of our proposition follows immediately from Lemma 6.35. In the second case we deduce from (12) in Section 7 that for any line a there exist elements u, v in S such that $a = uv$. Lemma 6.36 then implies $a^\perp = S(uv)$, as required.

If $E(G, S)$ is an elliptic plane, then Proposition 6.10 implies that all points in $E(G, S)$ are completely connected. Let a be any line in S which is not a central line; then we choose any point A not on the line a. By Proposition 5.11 A and $a^\perp$ are connected by a line b. Similarly, we choose a point B not on a or on b, and find a line c in $B \cap a^\perp$. Clearly, $b \neq c$, and by Theorem 5.4 $a^\perp \subset S(bc)$. Moreover, we have $S(bc)^a = S(bc)$, and therefore $S(bc) \subset a^\perp$ or $a \in S(bc)$. In the first case we obtain $S(bc) = a^\perp$, as required, and in the second case we choose for any line $x \neq a$ in $S(bc)$ a point $C \neq S(bc)$ on x. Then C, $a^\perp$ are connected by a line d by Proposition 5.11, and from $x, d \in C$, $S(bc)$ and $C \neq S(bc)$ we deduce $x = d \in a^\perp$. Hence we have $S(bc) = a^\perp \cup \{a\}$, as required. ∎

From the end of our proof for Proposition 6.37 we note for use later on:

(A)   *In an elliptic plane we have* $|a^\perp| \geq 2$ *for all lines* a.

Let a be any line in a complete S-group plane which is not a central line. Then the point $a^\perp$ or $a^\perp \cup \{a\}$ is denoted by S_a. If $|a^\perp| \geq 2$, then S_a is a polar point. Thus for any line a in a complete S-group plane which is not a central line there is at most one polar point (and in a Strubecker plane there are no polar points).

By our Main Theorem 6.1 any complete S-group plane $E(G, S)$ can be considered to be a suitable incidence structure $I(V, Q)$ over its coordinate field K. We wish to interpret orthogonality and polar points in $E(G, S)$ in terms of the metric vector space (V, Q). We show:

LEMMA 6.38. *Let* $E(G, S)$ *be a complete S-group plane considered as an incidence structure* $I(V, Q)$ *over its coordinate field* K. *Then for any two distinct lines* a, b *in* S *we have**

$$a \perp b \Leftrightarrow g_Q(a, b) = 0$$

---

* We write $g_Q(M, N) = 0$ for $M \subset V$ and $N \subset V$ if $g_Q(X, Y) = 0$ for all X in M and all Y in N, and note that a, b are sets of vectors in V.

PROOF. If $E(G, S)$ is one of the exceptional complete S-group planes, then our statement is true. Now let $E(G, S) \neq E(G_0, S_0), E(G_1, S_1)$. By our remarks above, we may assume $E(G, S)$ is not a Strubecker plane. From Theorem 6.31 we deduce that (in the notation introduced in Chapters 1, 3, and 4)

$$x^* = \bar{\sigma}_X \qquad \forall_X = \langle X \rangle \quad in \quad S. \qquad (\circ)$$

Thus we have the following equivalent statements for any two distinct lines $a = \langle A \rangle$ and $b = \langle B \rangle$ in S [compare the definition of σ_x in $S(V, Q)$]: $a \perp b \Leftrightarrow ab^* = a \Leftrightarrow \langle A \rangle \bar{\sigma}_B = \langle A \rangle \Leftrightarrow$ there is an element s in K such that $A\sigma_B = sA \Leftrightarrow (s + 1)A = (A \circ B)B \Leftrightarrow$ (since A, B are linearly independent) $A \circ B = 0 \Leftrightarrow g_Q(A, B) = 0 \Leftrightarrow g_Q(a, b) = 0$.

This proves our Lemma. ∎

11. CHARACTERISTIC OF A COMPLETE S-GROUP PLANE

To separate the case of characteristic $= 2$ from the case of characteristic $\neq 2$ we introduce a new axiom, the *anti-Fano axiom*. If a, b, c, d is a quadrilateral in an incidence structure (L, κ) and if $L(b, c)$, $L(a, d)$ are connected by a line e and if $L(a, c)$, $L(b, d)$ are connected by a line f, then the 6-tuple $\{a, b, c, d; e, f\}$ is called a *complete quadrilateral*. The points $L(a, b)$, $L(c, d)$, $L(e, f)$ are called the *diagonal points* of the complete quadrilateral.

Anti-Fano AXIOM. *There is a complete quadrilateral* $\{a, b, c, d; e, f\}$ *in* $E(G, S)$ *such that the points* $S(ab)$, $S(cd)$, $S(ef)$ *are collinear*.

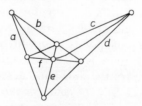

Let $E(G, S)$ be any complete S-group plane. Then we call $E(G, S)$ a *complete S-group plane of char 2* if the anti-Fano axiom holds or if $E(G, S)$ is isomorphic to one of the five incidence structures $I(V, Q)$ over $GF(2)$, or over $GF(4)$ in the case of Ind $(V, Q) = 1$. Otherwise $E(G, S)$ is called a *complete S-group plane of characteristic* $\neq 2$.

THEOREM 6.39. *A complete S-group plane* $E(G, S)$ *is of char 2 or of char* $\neq 2$ *if and only if its coordinate field is of char 2 or of char* $\neq 2$, *respectively*.

PROOF.   If $E(G, S)$ is one of the five exceptional cases, it is isomorphic to $I(V, Q)$ over $GF(2)$ or over $GF(4)$ in the case $\text{Ind}(V, Q) = 1$; our assertion then holds by definition.

Now suppose the complete S-group plane $E(G, S)$ is not one of the five exceptional cases. By our Main Theorem 6.1 we may consider $E(G, S)$ a suitable incidence structure $I(V, Q)$ over its coordinate field K. If $K = GF(3)$ and $\text{Ind}(V, Q) = 1$ and $\dim \text{rad}(V, Q) = 1$, then, obviously, the anti-Fano axiom does not hold. Hence $\text{char } E(G, S) \neq 2$ and $\text{char } K \neq 2$, as required. In all other cases there is a complete quadrilateral $\{a, b, c, d; e, f\}$ in $E(G, S)$ such that two of the three diagonal points $S(ab)$, $S(cd)$, $S(ef)$ are connected by a line, since $\text{Ind}(V, Q) = 0$ and therefore $I(V, Q)$ is a projective incidence structure or an affine incidence structure or a star-complement (compare Proposition 2.4), or we have $\text{Ind}(V, Q) = 1$ and $|K| \geq 5$ or $K = GF(3)$ and $\dim \text{rad}(V, Q) = 0$ (in the last case it is easily seen that there are four ends and we can choose a, b, c, d; e, f as the six lines connecting the ends).

By the Appendix the coordinate field K is of char 2 if and only if the projective coordinate incidence structure $I(V)$ is of char 2, and this is the case if and only if there is a complete quadrilateral with collinear diagonal points. If $I(V)$ is of char 2, then any complete quadrilateral has collinear diagonal points. Thus we have the following implications: $E(G, S)$ is of char $2 \Rightarrow$ the anti-Fano axiom holds $\Rightarrow$ there is a complete quadrilateral with collinear diagonal points in $E(G, S) \Rightarrow$ there is a complete quadrilateral with collinear diagonal points in $I(V) \Rightarrow$ char of $I(V)$ is $2 \Rightarrow$ char $K = 2$. Conversely, we see successively: char $K = 2 \Rightarrow$ char of $I(V)$ is $2 \Rightarrow$ any complete quadrilateral in $I(V)$ has collinear diagonal points $\Rightarrow$ the complete quadrilateral $\{a, b, c, d; e, f\}$ has collinear diagonal points $\Rightarrow$ $\{a, b, c, d; e, f\}$ is a complete quadrilateral in $E(G, S)$ and the line in S connecting two of the diagonal points $S(ab)$, $S(cd)$, $S(ef)$ must go through the last of these points $\Rightarrow$ there is a complete quadrilateral with collinear diagonal points in $E(G, S) \Rightarrow$ the anti-Fano axiom holds $\Rightarrow$ $E(G, S)$ is of char 2. Thus we have $E(G, S)$ is of char $2 \Leftrightarrow$ the coordinate field K is of char 2.

This proves the theorem.   ■

PROPOSITION 6.40.   *Any Strubecker plane is of char* $\neq 2$.

PROOF.   Let $E(G, S)$ denote any Strubecker plane. By Theorem 6.21 we may consider $E(G, S)$ a Strubecker coordinate plane $I(V, Q)$ over its coordinate field K. We have $\text{Ind}(V, Q) = 0$ and $\dim \text{rad}(V, Q) = 2$ and $\dim V^{\perp} \leq 2$. Hence $\dim V^{\perp} = 2$, and from Lemma 1.9 we see that char $K \neq 2$, and therefore Theorem 6.39 implies that $E(G, S)$ is of characteristic $\neq 2$. ■

PROPOSITION 6.41. *For a complete S-group plane* E(G, S) *that is not a Strubecker plane the following statements are equivalent:*

(a) E(G, S) *is of characteristic* $\neq 2$:
(b) *For all lines* a *in S we have* $a \notin Z(G)$ *and* $a \notin S_a$.
(c) *There is a line* a *such that* $a \notin Z(G)$ *and* $a \notin S_a$.
(d) *There are lines* a, b, c *such that* $a \notin Z(G)$ *and* $b, c \in a^\perp$ *and* $abc \notin J$.

PROOF. We may consider the S-group plane $E(G, S)$ an incidence structure $I(V, Q)$ over its coordinate field K (compare the Main Theorem 6.1).

(a) $\Rightarrow$ (b): Let $E(G, S)$ be of char $\neq 2$. Then Theorem 6.39 implies char $K \neq 2$. Let $a = \langle A \rangle$ be any line in $S(= L(V, Q))$. Then $Q(A) \neq 0$, and therefore by equation 1 in Chapter 1, Section 1 $g_Q(A, A) \neq 0$ and thus $\langle A \rangle^\perp \neq V$ and dim $\langle A \rangle^\perp = 2$ by Lemma 1.1. This implies $\bar{\sigma}_A \neq 1$, and therefore by $(\circ)$ $\langle A \rangle^* \neq 1$ and a is not a central line. Moreover, any nonisotropic line in $\langle A \rangle^\perp$ is distinct from a,* hence by Lemma 6.38 in $a^\perp$. Since char $K \neq 2$, we have $|K| \geq 3$, and therefore the point $\langle A \rangle^\perp$ in $I(V)$ contains at least four lines, hence $a^\perp \subset \langle A \rangle^\perp$ at least two lines, say b, c. If $a \in S_a$, then a, b, c are concurrent, hence $a \in \langle A \rangle^\perp$ and $A \in \langle A \rangle^\perp$, a contradiction. Thus we have $a \notin S_a$.

(b) implies (c), trivially.

(c) $\Rightarrow$ (d): Let a be any line such that $a \notin Z(G)$ and $a \notin S_a$. Then we see from Proposition 6.37 that $S_a = a^\perp$. For any two distinct lines b, c in the point S_a we then have b, $c \in a^\perp$ and $abc \notin J$.

(d) $\Rightarrow$ (a): Let a, b, c be three lines such that $a \notin Z(G)$ and $b, c \in a^\perp$ and $abc \notin J$. Then $b \neq c$ and $a \notin S(bc)$ and $S(bc)^a = S(bc)$, hence $S(bc) = a^\perp$ by Theorem 5.4. From Lemma 6.38 we deduce $S(bc) \subset \langle A \rangle^\perp$ for $\langle A \rangle := a$. Assuming $A \in \langle A \rangle^\perp$, we would obtain a, b, c are concurrent in $I(V, Q)$ [since dim $\langle A \rangle^\perp = 2$ in consequence of $a \notin Z(G)$]; thus a, b ,c would be concurrent in $E(G, S)$, in contradiction to $a \notin S(bc)$. Hence $A \notin \langle A \rangle^\perp$ and therefore $2 = g_Q(A, A)Q(A)^{-1} \neq 0$. This yields char $K \neq 2$, and by Theorem 6.39 $E(G, S)$ is of char $\neq 2$. ■

---

* By our identification of $E(G, S)$ and $I(V, Q)$ we must pay attention to the fact that for a in $S = L(V, Q)$ there are two distinct meanings of $a^\perp$. Since a is a one-dimensional subspace $\langle A \rangle$ of V, $a^\perp$ denotes the set of perpendiculars to a in $E(G, S)$ as well as the set of orthogonal vectors to $\langle A \rangle$. To distinguish the two cases we write $a^\perp$ only for the set of perpendiculars to $a = \langle A \rangle$ in $E(G, S)$, and $\langle A \rangle^\perp$ for the orthogonal subspace to $\langle A \rangle = a$.

Keeping in mind that in a complete S-group plane $E(G, S) \neq E(G_0, S_0)$ there exists at most one central line, we see that Proposition 6.41 yields immediately:

PROPOSITION 6.42. *For any complete S-group plane* $E(G, S) \neq E(G_0, S_0)$ *that is not a Strubecker plane the following statements are equivalent:*

(a) $E(G, S)$ *is of char 2.*
(b)   *There is a line* a *such that* $a \notin Z(G)$ *and* $a \in S_a$.
(c)   *For all lines* a *not in* $Z(G)$ *we have* $a \in S_a$.

A more specialized characterization of the two cases of the characteristic of a complete S-group plane can be given by separating into the four cases of the elliptic, Euclidean, hyperbolic-metric, and Minkowskian planes. We recall that the Strubecker planes are always of char $\neq 2$.

PROPOSITION 6.43.   *For an elliptic or hyperbolic-metric plane* $E(G, S)$ *we have:*

(a) $E(G, S)$ *is of char* $2 \Leftrightarrow$ *there is a central line.*
(b) $E(G, S)$ *is of char* $\neq 2 \Leftrightarrow$ *there is a polar triangle.*

PROOF.   If $E(G, S)$ is the exceptional plane $E(G_1, S_1)$, then there is a central line and the plane is of char 2.

Thus we may assume $E(G, S) \neq E(G_1, S_1)$. Then there are four lines, no three of them concurrent, and Corollary 4.6 yields that in $E(G, S)$ there cannot be a central line and a polar triangle. However, we can show:

(1)   *If* $E(G, S)$ *is an elliptic or hyperbolic-metric plane, then there exists a central line or a polar triangle.*

This follows from (11) in Section 7 if $E(G, S)$ is a hyperbolic-metric plane.

Now let $E(G, S)$ be an elliptic plane. Then all the points of $E(G, S)$ are completely connected by Proposition 6.10. We assume there is no central line. Then S_a is defined for all lines a in S. If $a \in S_a$, then we choose any line b not in S_a. From Proposition 5.11 we see that there is a line c in $S_a \cap b^\perp$. We have $c \neq a$ and therefore $c \in a^\perp$, otherwise $b \in a^\perp$ and $b \notin S_a$, which is a contradiction. By (A), there is a line $d \neq c$ in $a^\perp \subset S_a$. Then we deduce from Lemma 6.34 that $cd = dc$, and therefore a, b, $d \in c^\perp$ and $abd \notin J$. This implies $c \in Z(G)$ by Theorem 5.4, in contradiction to our assumption. Hence $a \notin S_a$ and $S_a = a^\perp$ for all lines in $E(G, S)$. We choose any line a and any line b in $a^\perp$. Then $a^\perp \neq b^\perp$ (since $b \in a^\perp$ and $b \notin b^\perp$), and there is a line c connecting the points $a^\perp$ and $b^\perp$. From $ab, ac, bc \in J$ we deduce $abc = 1$, since $abc \in J$ would imply $a^\perp = S(bc) = S(ac) = b^\perp$. Hence a, b, c is a polar triangle, and (1) is proved.

From (1) we see successively, applying Lemma 6.35 and 6.36 and Proposition 6.41 and 6.42:

$E(G, S)$ is of char $2 \overset{6.42}{\Leftrightarrow} a \in S_a$ for all a in $S \setminus Z(G) \overset{6.36}{\Rightarrow}$ there does not exist a polar triangle $\overset{(1)}{\Rightarrow}$ there exists a central line.

Conversely: If there is a central line $\overset{6.35}{\Rightarrow} a \in S_a$ for all a in $S \setminus Z(G) \overset{6.42}{\Rightarrow}$ $E(G, S)$ is of char 2.

Thus (a) is true. From this and by (1) we conclude:

$E(G, S)$ is of char $\neq 2 \Leftrightarrow$ there is no central line $\overset{(1)}{\Leftrightarrow}$ there is a polar triangle.

This proves (b). ∎

PROPOSITION 6.44. *For any line in a Euclidean or Minkowskian plane* $E(G, S) \neq E(G_0, S_0)$ *of char 2 we have* $\Pi_a = S_a$ *(if* Π_a *denotes the pencil of the lines parallel to* a). *Thus for any two distinct lines* a, b:

$$a \parallel b \Leftrightarrow a \perp b \tag{1}$$

PROOF. Let a be any line in $E(G, S)$. Then a is not a central line by Proposition 6.7(a). Then S_a is defined and is a 2-Δ-connected point by Proposition 6.8. From ($\circ$) in Section 5 or 8 we see that Π_a is a 2-Δ-connected point. Since $E(G, S)$ is of char 2, we obtain from Proposition 6.42 that $a \in S_a$. Obviously, $a \in \Pi_a$. Since Proposition 6.8(c) yields that any two distinct 2-Δ-connected points are not connected by a line, we deduce from $a \in \Pi_a$, S_a that $\Pi_a = S_a$. Since $S_a = a^\perp \cup \{a\}$ by Proposition 6.37, we arrive immediately at equation 1. ∎

7

S-GROUP PLANES
WITH COMPLETELY
CONNECTED POINTS

According to Chapter 6, we have treated an axiomatic-synthetic approach to "full" metric planes over an arbitrary field. However, in absolute geometry we may be interested also in substructures of these full metric planes. In fact, there are many interesting examples of such substructures, and many interesting questions and problems concerning these substructures. To arrive at substructures of complete metric planes, we have to weaken the assumptions we made in Chapter 6. However, in the process of weakening our assumptions we must be careful: we must remain sure that the given incidence structure can be embedded in a complete metric plane up to isomorphisms. In recent developments of absolute geometry we find various conditions suggested for this purpose. However, there are no "natural" assumptions that are sufficient to establish an embedding of the given incidence structure. In a certain sense we give here an axiomatic approach, which is most general at the present stage.

Our assumptions can be formulated in terms of S-group planes with Δ-connected points. Thus our treatment in this section undertakes the investigation of all the S-group planes with Δ-connected points.

1. DEFINITION OF S-GROUP PLANES WITH COMPLETELY CONNECTED POINTS, AND THE MAIN THEOREM

An S-group plane is called an *S-group plane with completely connected points* if there exists at least a quadrilateral and if there exists at least one completely connected point $S(ab)$ such that $ab \neq ba$.

Let $E(G, S)$ be any S-group plane with completely connected points. Then by Corollary 5.3 and Theorem 4.11 $(S, \kappa(G, S), \Phi(G, S))$ is an S-plane, which we may also call the group plane of the S-group (G, S).

To formulate our main theorem on S-group planes with completely connected points, we introduce the following concepts:

Let (L, κ) and (L', κ') be two incidence structures. Then (L, κ) is said to be *embeddable* in (L', κ') if (L, κ) is isomorphic to a suitable substructure of (L', κ'). We say (L, κ) is *properly embeddable* in (L', κ') if the isomorphism of (L, κ) onto the substructure of (L', κ') maps the set of points of (L, κ) onto the set of points of (L', κ').

Now let (L, κ, Φ) and (L', κ', Φ') be any two S-planes; then (L, κ, Φ) is called *embeddable* in (L', κ', Φ') if, up to isomorphisms, (L, κ, Φ) is an S-subplane of (L', κ', Φ'). Clearly, we say (L, κ, Φ) is *properly embeddable* in (L', κ', Φ') if (L, κ) is properly embeddable in (L', κ').

It is obvious how we have to define an embeddability of an incidence structure with reflections (L, κ, φ) into an incidence structure with reflections (L', κ', φ'). Then this general definition would contain the aforementioned definition as well as the definition given in Chapter 5, Section 7.

MAIN THEOREM 7.1. *Any S-group plane with completely connected points is properly embeddable in a complete metric plane.*

The complete metric plane in which the given S-group plane $E(G, S)$ with completely connected points can be properly embedded is called the *ideal plane* of $E(G, S)$.

In the following we elaborate a proof of this theorem.

If the given S-group plane with completely connected points is finite, then our assertion follows immediately from Theorem 8.5, which is proved in Chapter 8 independently of our discussions in this section. Thus we may restrict ourselves to the case of infinite S-group planes with completely connected points.

This restriction is necessary, since our proof given in the following is essentially based on an application of Theorem 5.13, the theorem of Desargues. However, the proof of that theorem assumes that, if a line contains at least one completely connected point, then the line is contained in at least six completely connected points. In general, this is not true for any S-group plane with completely connected points. However, we show in the next paragraph that only a few examples of S-group planes with completely connected points fail to satisfy the aforementioned condition, and that all these examples are finite.

2. PROPER POINTS ON A LINE

Let $E(G, S)$ be any S-group plane with completely connected points. For the sake of brevity, we call any completely connected point $S(ab)$ for which $ab \neq ba$ a *proper point.* Let A denote a proper point in $E(G, S)$. Then we prove

(1) A *does not contain three distinct lines* a, b, c *such that* $a, b \in c^{\perp}$.

Suppose the contrary. Since there exists a quadrilateral, we can choose a line d such that $d \notin A$. By Proposition 5.11 there exists a line z in $A \cap d^{\perp}$. Then $d, acz, zcb \in z^{\perp}$ and $d(acz)(zcb) = dab \notin S$. Theorem 5.4 then yields that z is a central line. Hence for any two lines x, y in A we have $zx = xz$ and $zy = yz$ and $xyz = zyx$ and therefore $xy = yx$, in contradiction to the fact that A is a proper point.

(2) $A^x = A^y$ *and* $x, y \notin A$ *imply* $xy = yx$.

Since A is completely connected, there exists a line s in A such that $sxy \in S$. Then we have $A^{sxy} = A$ and $sxy \notin A$ or $sxy \in A$. In the first case we obtain $A \subset (sxy)^{\perp}$ and therefore $xy = s(sxy) \in J$. In the second case we obtain $x = y$, since $x \neq y$ would imply $s, sxy \in A$, $S(xy)$ and $A \neq S(xy)$ and therefore $s = sxy$ and $x = y$, a contradiction. Thus in both cases we deduce $xy = yx$, as required.

(3) $A \subset x^{\perp}$ *and* $A \subset y^{\perp}$ *imply* $x = y$.

From $A \subset x^{\perp}$ and $A \subset y^{\perp}$ we see that $A^x = A = A^y$ and $x, y \notin A$. Hence $x, y \in z^{\perp}$ for any z in A, and by (2) $xy = yx$. If $x \neq y$, then $xyz = 1$ or $z \in S(xy) \neq A$ for any line z in A, which yields $|A| \leq 2$ and $uv = vu$ for all u, v in A, in contradiction to the fact that A is a proper point. Hence $x = y$.

(4) A^S *contains at least three noncollinear points.*

Suppose the contrary and assume $z \in A^x$ for all x. Then $x \in z^{\perp}$ for all x not in A. Since A is a proper point, we can find a line s in A such that $sz \neq zs$. Then $s^x \in A$ for all x not in A (as otherwise we would obtain $s^x \in z^{\perp}$ and therefore $s^x z = zs^x$ and thus $sz = zs$). This implies $s, s^x \in A$, $S(sx)$ and $A \neq S(sx)$ and therefore $s^x = s$. Hence $A^x = S(sz)^x = S(sz) = A$ for all x not in A. By the existence of a quadrilateral this implies a contradiction to (3).

(5)   *Let* B *be any proper point* $\neq A$. *Then any three distinct points in* A^B *are noncollinear, or there exists a line* z *in* A *such that* $B \subset z^{\perp}$.

We assume that there are three distinct collinear points in A^B. Since A and B are connected, we may assume that there are lines x, y, z such that $z \in A, A^x, A^y$ and $x, y \in B$ and A, A^x, A^y are mutually distinct. Then $x, y \in z^\perp$, and by (1) $z \notin B$. Hence $B = S(xy) \subset z^\perp$, as required.

(6) *There exist two proper points* B, C *in* A^S *such that* B *does not contain a line z with* $C \subset z^\perp$, *provided* $|A| > 4$.

By (4) there exist three noncollinear points in A^S, say C, D, E. From (3) we see that there is at most one line z such that $C \subset z^\perp$. If there exists such a line and if $z \in D$, E (only in this case our assertion cannot be satisfied), we can find a line s in E such that $z \notin D^s$ and $D^s \neq C$: Since E is a proper point, there exists a line t in E such that $tz \neq zt$ and therefore $z \notin D^t$. If $D^t \neq C$, then $B := D^t$ fulfills our assertion. If $D^t = C$, then E contains a line $s \neq t$, r, z, trz (if r denotes the line in $C \cap E$), since $|E| = |A| > 4$. From (1) we see that $sz \neq zs$ and therefore $z \notin D^s$, and from (2) we deduce that $D^s \neq C$, and $B := D^s$ satisfies our assertion.

(7) *If* $|A| > 4$, *then for any line g we have*

$$|g^\perp| \geq \tfrac{1}{2}|A| - 1$$

By (6) there exist two distinct proper points B, C in A^S such that for any line z we have $z \notin B$ or $C \not\subset z^\perp$. Then (5) yields that any three distinct points in B^C are noncollinear. Hence g is contained in at most two points in B^C, and any point in B^C not on g is connected with $g^\perp$ by Proposition 5.11. Any joining line from a point in B^C to the set $g^\perp$ is contained in at most two points of B^C by (5). Hence from $|B^C| = |C| = |A|$ we can deduce our assertion.

(8) *If the line g is contained in a proper point* B, *and if* $|A| > 4$, *then the number of proper points on g is greater than or equal* $\tfrac{1}{4}|A|$.

We put

$$E := \{B^x \mid x \in g^\perp\}$$

Obviously, E contains only proper points on g. By (1) at most one of the lines in $g^\perp$ is a line in B. Hence $|g^\perp| - 1$ lines in $g^\perp$ are not in B. If $B^x = B^y$ for $x, y \in g^\perp \backslash B$ and $x \neq y$, then we deduce, denoting the joining line in $B \cap S(xy)$ by s, $B^{sxy} = B$. Now put $z := sxy$.

Obviously, we have $z \notin B$ (as otherwise we would obtain $x = y$) and therefore $B \subset z^\perp$. By (3) the line z is uniquely determined. Thus given any line x in $g^\perp \backslash B$, there is at most one line $y \neq x$ in $g^\perp \backslash B$ such that $B^x = B^y$. Hence $|E| \geq \frac{1}{2}(|g^\perp| - 1) + 1 \geq \frac{1}{4}|A|$, as required.

(9)   *If* $E(G, S)$ *is infinite, then* $|A|$ *is infinite.*

If $E(G, S)$ is infinite, then S is infinite. Suppose $|A|$ is finite; then $S \backslash A$ is infinite. There exists a line b not in A. Since any point on b is connected with A, and since any two lines joining A with a point on b are distinct, we have $|b| = |A|$ and therefore $|b|$ is finite.* Hence one of the points on b has an infinite number of lines, say B. Choose any line c not in B. Then $|c|$ is infinite, since all the lines in B intersect c in distinct points. By (4) there is a point C in A^S not on c. Since C is connected with all the points on c by distinct lines, $|C|$ is infinite. However, it is clear that $|C| = |A|$, since $C \in A^S$, and therefore $|C|$ is finite, a contradiction. This proves (9).

From (9) and (8) we deduce immediately:

PROPOSITION 7.2.   *Let* $E(G, S)$ *be an infinite S-group plane with completely connected points. If a line is contained in a proper point, then it is contained in an infinite number of proper points.*

This completes our restriction to the infinite case.

3.   GERMS OF PERSPECTIVE COLLINEATIONS

In the following let $E(G, S)$ be any fixed S-group plane with completely connected points. According to Section 2 we may assume that $E(G, S)$ is infinite. Thus by Proposition 7.2 we are sure that there are an infinite number of proper points on a line, as long as this line is contained in at least one proper point.

By (8) we may restrict this general assumption to the assumption that the proper point O, chosen fixed for the following, contains at least 25 lines, since this implies that any proper point contains at least 25 lines and therefore any line that is contained in at least one proper point is contained in at least six proper points.

To construct the ideal plane of $E(G, S)$ we only have to introduce new lines (ideal lines). Any ideal line is defined as a suitable set of points. To have a uniform definition of lines in the ideal plane of $E(G, S)$ we wish to describe also the lines in $E(G, S)$ as sets of points. This can be done in the following way: Let a be any line in S; then we put

$$l(a) := \{X \mid X \text{ is a point on } a\}$$

---

* We recall that $|x|$ for a line x denotes the number of points on that line.

and call $l(a)$ an *ideal line*. Thus we have to distinguish between a line a in $E(G, S)$ and the ideal line $l(a)$ in the ideal plane of $E(G, S)$, as defined in the following. In general, we see that there exist ideal lines that are not ideal lines of the type $l(a)$. Obviously, the map $a \mapsto l(a)$ is an injective map of S into the set of ideal lines.

We call any ideal line $l(a)$ containing at least one proper point a *proper ideal line*. By Proposition 7.2 any proper ideal line is contained in an infinite number of proper points provided $E(G, S)$ is infinite.

In the following we let O be some fixed proper point and A, B two points $\neq O$, connected with O by the same line.

Given any proper ideal line $l(o)$ such that $o \notin O$ we wish to define a germ of a perspective collineation with the center O and the axis o [or say $l(o)$] mapping A onto B. To do this we choose any two distinct completely connected points U, V on o and put

$$P(V) := \{X \mid (O, V) \notin X \quad \text{or} \quad X = O \quad \text{or} \quad X = V\}$$

We consider the map

$$\omega_{UV} : \begin{cases} P(V) \to P(V) \\ X \mapsto \begin{cases} X \text{ if } o \in X \text{ or } X = O \\ (X, O) \cdot (V, ((U, B) \cdot (O, (V, X) \cdot (U, A)))) \end{cases} \end{cases}$$

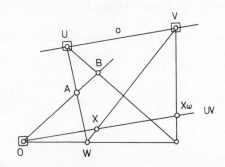

We recall that for any two distinct connected points X, Y the symbol (X, Y) denotes the joining line of X and Y, and that for any two distinct lines a, b in $E(G, S)$ the symbol $a \cdot b$ denotes the point $S(ab)$.

Obviously, ω_{UV} is a bijective map of $P(V)$ onto itself. Moreover, we have

$$O, X, X\omega_{UV} \quad \text{are collinear} \quad \forall X \quad \text{in} \quad P(V) \tag{1}$$

$$l(x)\omega_{UV} = l(x) \, \forall x \neq (O, V) \quad \text{in} \quad O \tag{2}$$

Moreover, we can show for any completely connected point $V' \neq U$ on o:

(10) $X\omega_{UV'} = X\omega_{UV} \ \forall X \in P(V) \cap P(V')$.

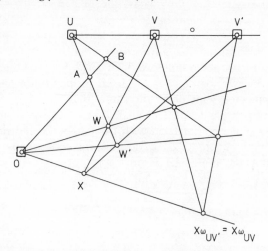

We may assume $V \neq V'$ and $X \neq O$ and $o \notin X$. If $X = (U, A) \cdot (V, X)$, our assertion holds trivially, since in this case also we have $X = (U, A) \cdot (V', X)$. In all other cases the assumptions of Theorem 5.13 (Desargues) are satisfied. From the assertion of this theorem we see that $X\omega_{UV'} = X\omega_{UV}$.

Similarly, we prove for any completely connected point $U' \neq V$ on o:

(10') $X\omega_{UV} = X\omega_{U'V} \ \forall X \in P(V)$.

We may assume $U \neq U'$ and $X \neq O$ and $o \notin X$. If $A = (U, A) \cdot (V, X)$, then

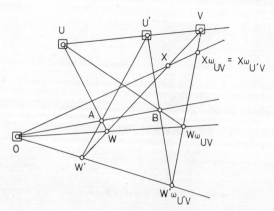

we have $X\omega_{UV} = X\omega_{U'V}$, trivially. In all other cases we may apply Theorem 5.13 and obtain our assertion $X\omega_{UV} = X\omega_{U'V}$.

We now want to extend the map ω_{UV}, which is only defined on $P(V)$, to a map ω on the set P of all points. For this we choose three distinct fixed completely connected points U, V, V' on o and define the map ω of the set P of all points on itself by putting

$$X\omega := \begin{cases} X\omega_{UV} & \text{if} \quad X \in P(V) \\ X\omega_{UV'} & \text{if} \quad (O, V) \in X \end{cases}$$

Then we can show

(11) *For any two distinct completely connected points* U', V'' *on* o *such that* $U' \neq V''$ *we have*

$$X\omega = X\omega_{U'V''} \qquad \forall X \in P(V'')$$

If $X \in P(V) \cap P(V'')$, then we see by definition and from (10) and (10'): $X\omega = X\omega_{UV} = X\omega_{UV''} = X\omega_{U'V''}$. If $X \in P(V'')$, but $X \notin P(V)$, then we have $(O, V) \in X$ and therefore $(O, V') \notin X$ and $X \in P(V')$. By definition and from (10) and (10') we obtain $X\omega = X\omega_{UV'} = X\omega_{UV''} = X\omega_{U'V''}$. This proves (11).

Statement (11) shows that the map ω is independent of the choice of the two completely connected points U and V on o. We call ω the *germ of a perspective collineation*, having O as a center and o as an axis and mapping A onto B. Sometimes we denote this map by ω^{AB}. The set of all germs of perspective collineations with center O and axis o is denoted by $P(O, o)$. For any proper point O we define

$$K(O) := \bigcup_{\substack{o \notin O \\ l(o)\text{ proper line}}} P(O, o)$$

We next show

(12) *Let* $\omega = \omega^{AB} \in P(O, o)$ *and let* X *be any point not on* o. *Then* $x \in X$ *and* $x' \in X\omega$, $S(ox)$ *imply* $l(x)\omega = l(x')$.

If $x \in O$, then by our remark above, $l(x)\omega = l(x)$ and therefore $x \in X\omega$, $S(ox)$. Obviously, we have $X\omega \neq S(ox)$ (as otherwise $o \in X\omega$ and therefore $o \in X$) and $x, x' \in X\omega$, $S(ox)$. This yields $x = x'$ and therefore $l(x)\omega = l(x) = l(x')$.

Now let $x \notin O$. Then, in particular, $X \neq O$. Let Y be any point on x and $o \notin Y$. Then we choose two distinct completely connected points $U, V \neq S(xx')$ on o, such that $X, Y \in P(V)$ (this is possible, if $|O| > 16$).

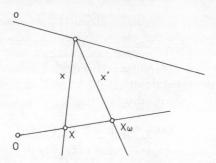

We put $X':=X\omega=X\omega_{UV}$; $Y':=Y\omega=Y\omega_{UV}$; $P_2:=(U,A)\cdot(V,X)$; $Q_2:=(U,B)\cdot(V,X')$; $P_2':=(U,A)\cdot(V,Y)$; $Q_2':=(U,B)\cdot(V,Y')$; $P_3:=x\cdot(U,A)$; $Q_3:=x'\cdot(O,P_3)$; $Q_2'':=(U,Q_3)\cdot(O,P_2)$. Then the assumptions of Theorem 5.13 for the triangles X, P_2, P_3 and X', Q_2'', Q_3

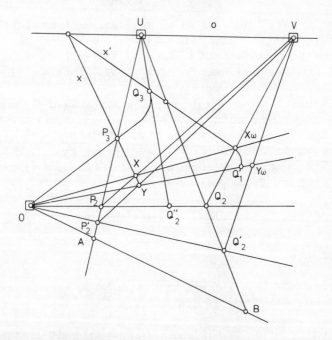

are satisfied; in particular, O, U, V are completely connected points. By Theorem 5.13 we obtain that X', Q_2'', V are collinear. This yields $Q_2=Q_2''$ and therefore $Q_3=S(x'y)$ if $y:=(U,B)$. Furthermore, the assumptions of Theorem 5.13 are fulfilled for the triangles Y, P_2', P_3 and $Q_1':=x'\cdot(O,Y')$, Q_2', Q_3. Hence the points Q_1', Q_2', V are collinear. This yields $Q_1'=Y\omega$ and therefore $Y\omega\in l(x')$. Thus we have proved that

$Y \in l(x)$ implies $Y\omega \in l(x')$, and hence we are ready, since the inverse inclusion $l(x') \subset l(x)\omega$ follows using the mapping ω^{-1}.

(13) $\omega \in P(O, o)$ *implies* $\omega^{-1} \in P(O, o)$.

Let $\omega \in P(O, o)$ and A be any point $\neq O$ not on o. We put $B := A\omega$. By our construction we have $\omega^{-1} = \omega^{BA}$ and therefore $\omega^{-1} \in P(O, o)$.

(14)   *Let $\omega \in P(O, o)$ and $\omega' \in P(O, o')$ and o, o' be two lines in a proper point U. Then there exists a line o'' in U such that $o'' \notin O$ and $\omega\omega' \in P(O, o'')$ or $o'' \in O$ and $\omega\omega'$ fixes any proper ideal line $l(x)$ such that $x \in O$ and at least three points on o''.*

If $o = o'$, then we choose any point $A \neq O$ not on o and consider the germ $\omega'' := \omega^{AA\omega\omega'}$ of a perspective collineation in $P(O, o)$. By construction, and by the independence of the elements of $P(O, o)$ on the choice of the completely connected points U and V on o, we immediately obtain that $\omega'' = \omega\omega'$. Thus we may assume $o \neq o'$.

We choose a completely connected point V on o and a completely connected point V' on o' such that $V, V' \neq U$ and $(V, V') \notin O$. Further, we choose a point C on (V, V') such that $C \neq V, V'$. Then we put $A := C\omega^{-1}$ and $B := (A, V) \cdot (A\omega\omega', V')$ and $o'' := (U, B)$.

If $o'' \in O$, then we obtain $B\omega\omega' = B$, and o'' is contained in the three fixed points O, B, U. Obviously, $\omega\omega'$ fixes any proper ideal line $l(x)$ such that $x \in O$.

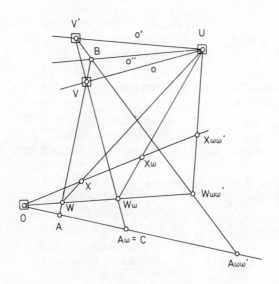

Now let $o'' \notin O$. To prove that $\omega\omega' \in P(O, o'')$ we consider the germ $\omega'' := \omega^{AA\omega\omega'}$ of a perspective collineation in $P(O, o'')$. Our aim is to prove $\omega'' = \omega\omega'$. Obviously, $\omega = \omega^{AA\omega}$ and $\omega' = \omega^{A\omega A\omega\omega'}$.

Let X be any point $\in P(U)$. Then we construct $X\omega$ and $X\omega\omega'$ as the points $X\omega_{VU}$ and $(X\omega)\omega_{V'U}$, respectively. From (12) we see that $(A, V)\omega'' = (A\omega\omega', V')$. We put $W := (U, X) \cdot (A, V)$ and obtain $W\omega'' = W\omega\omega'$, and by (12) $(U, X)\omega'' = (U, X\omega\omega')$. This yields $X\omega'' = X\omega\omega'$, as required.

To prove our assertion in the case $X \notin P(U)$, that is $(O, U) \in X$ and $X \neq O, U$, we first suppose that X is a completely connected point. Then

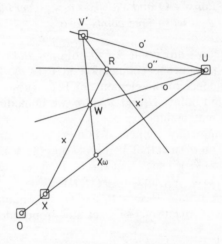

the joining line of V' and $X\omega$ does exist. We denote its intersection with the line o by W. Since X is completely connected, there exists a line x in $X \cap W$. Moreover, we can find a line x' in $V' \cap S(xo'')$. We put $R := S(xo'')$. By (12) we have $l(x)\omega = l((V', W))$. By the first part of our proof we obtain $R\omega\omega' = R\omega'' = R$. Combining the fact that $R\omega \in l(x)\omega$ with (12) lets us deduce that $l(x)\omega\omega' = l(x')$. We choose any point $Y \neq X, R$ on x and deduce from the first part of the proof $Y\omega\omega' = Y\omega''$. Then we see from (12) that $l(x)\omega'' = l(x')$ and therefore $l(x)\omega\omega' = l(x)\omega''$. Thus we have (O, U), $x' \in X\omega\omega'$, $X\omega''$ and $(O, U) \neq x'$ and therefore $X\omega'' = X\omega\omega'$, as required.

Finally, let X be any point $\neq O, U$ on (O, U). We choose a completely connected point $Y \neq O, U$ on (O, U). Then we put $\omega_1 := \omega^{YX} \in P(O, o)$. As proved in the first part, we have $\omega_1\omega \in P(O, o)$. Let us put $\omega_2 := \omega_1\omega''$. Then there exist lines o_1 and o_2 in U which are contained in at least three points fixed by $\omega_1\omega''$ and $\omega_2\omega'$, respectively. If $o_1 \in O$, then we interchange the point Y by another completely connected point $Y' \neq O, U$ on

(O, U). Then the corresponding line o_1' is not contained in O. By our general assumption there exist enough completely connected points on (O, U). Thus we may assume $o_1, o_2 \notin O$. Then we have proved that there are germs ω_3, ω_4 of perspective collineations in $P(O, o_1)$ and $P(O, o_2)$, respectively, such that $\omega_1\omega''$ and ω_3 coincide for all points in $P(U)$ and such that $\omega_2\omega'$ and ω_4 coincide for all points in $P(U)$. Thus for any point Z in $P(U)$ we obtain

$$Z\omega_4 = Z\omega_2\omega' = Z\omega_1\omega\omega' = Z\omega_1\omega'' = Z\omega_3$$

In particular, we deduce that $o_1 = o_2$ and therefore $\omega_3 = \omega_4$.

Since Y is completely connected, the arguments given above show that $Y\omega_2\omega' = Y\omega_4$ and $Y\omega_1\omega'' = Y\omega_3$. Thus we conclude $X\omega\omega' = Y\omega_1\omega\omega' = Y\omega_2\omega' = Y\omega_4 = Y\omega_3 = Y\omega_1\omega'' = X\omega''$, as required.

This completes the proof of (14).

From (13) and (14) we see that $P(O, o)$ is a group.

(15) *Let* A_1, A_2, A_3 *be any three distinct points. Then there exists a product* ω *of germs of perspective collineations in* $K(O)$ *for which the points* $A_i\omega (i = 1, 2, 3)$ *are three proper points or are collinear* ($i = 1, 2, 3$).

We may assume that O, A_1, A_2 are not collinear (as otherwise our assertion is true). Since O is a proper point, there exist two distinct proper points U, B_1 on (O, A_1). Let o' be any line $\neq (O, U)$ through U. Then there exists a germ of a perspective collineation ω' in $P(O, o')$ such

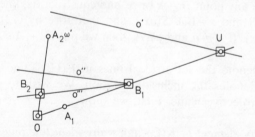

that $A_1\omega' = B_1$. Let o'' be any line $\neq (O, A_1)$ through B_1 such that $o'' \notin A_2\omega'$. Furthermore, we can choose a proper point $B_2 \neq O$ on $(O, A_2\omega')$ such that $o'' \notin B_2$. Then we consider the germ $\omega'' := \omega^{A_2\omega'B_2} \in P(O, o'')$. If $A_3 = O$, then $\omega := \omega'\omega''$ is a product of germs of perspective collineations that map A_1, A_2, A_3 onto the three proper points B_1, B_2, O. Thus we may assume that $A_3 \neq O$. Then we put $o''' := (B_1, B_2)$ and choose a proper point $B_3 \neq O$ on the line $(O, A_3\omega'\omega'')$ such that $o''' \notin B_3$. If

$o''' \in A_3\omega'\omega''$, then we have $o''' \in A_1\omega'\omega''$, $A_2\omega'\omega''$, $A_3\omega'\omega''$, and the assertion of (15) is true. If $o''' \notin A_3\omega'\omega''$, then we can consider the map $\omega''' := \omega^{A_3\omega'\omega''B_3} \in P(O, o''')$. For $\omega := \omega'\omega''\omega'''$ and $i = 1, 2, 3$ we obtain that $A_i\omega = B_i$ and B_1, B_2, B_3 are proper points, as required.

4. THE IDEAL INCIDENCE STRUCTURE OF AN S-GROUP PLANE WITH COMPLETELY CONNECTED POINTS

As above, let O be a fixed proper point and let $K(O)$ be the set of germs of perspective collineations in any set $P(O, o)$ such that $l(o)$ is a proper ideal line and $o \notin O$. As usual we denote the set of points in $E(G, S)$ by P.

A subset l of P is called an *ideal line* if there are a line x in S and an element ω in $K(O)$ such that $l = l(x)\omega$. Two ideal lines are identical if and only if the corresponding sets of points coincide. Any point A in P is called *incident* with the ideal line l if $A \in l$. Any set $l(x)$ for $x \in S$ is an ideal line in the sense of the definition just given, since the identity is an element in $K(O)$.

(16) *If an ideal line* l *contains a completely connected point* A, *then it is an ideal line of the form* l(x) *with* x ∈ S.

By definition there is an element x' in S and an element ω in $K(O)$ such that $l = l(x')\omega$. Let o denote the axis of ω. If $x' = o$, then obviously $l = l(x')$ and we are done. If $x' \neq o$ and $o \in A$, then we have $A = S(ox')$. We choose any point $B \neq A$ on x' and construct the line x in $A \cap B\omega$. We have $x' \in B$ and $x \in B\omega$, $S(ox')$, and therefore by (12) $l(x')\omega = l(x)$ and thus $l = l(x)$. If $x' \neq o$ and $o \notin A$, then we put $B := A\omega^{-1}$ and conclude as before.

Let L denote the set of ideal lines in $E(G, S)$ and put $\Pi := (P, L, \in)$, where $\in$ denotes the inclusion relation. We wish to prove that Π is a pappian projective plane. First, we show

(17) *Every element in* K(O) *maps any ideal line onto an ideal line.*

Let l be any ideal line and $\omega \in K(O)$. By definition there exist an element x in S and an element $\omega' \in K(O)$ such that $l = l(x)\omega'$. Let o, o' be the axes of ω and ω', respectively. We choose two proper points U, U' on o and o', respectively, such that $(U, U') \notin O$. Then we put $o'' := (U, U')$. Let A be any point $\neq O$ in l such that $o, o', o'' \notin A$ and B any proper point $\neq O$ on (O, A) with $o'' \notin B$. Then we consider the element $\omega_1 := \omega^{AB} \in P(O, o'')$. By (14) every choice of B yields a line containing at least three points fixed by $\omega'\omega_1$ and $\omega_1^{-1}\omega$. This line is in O for at most two choices of B. Hence we can choose such a proper point B on (O, A) that $\omega'\omega_1$ and $\omega_1^{-1}\omega$ are elements of $K(O)$ by (14). We put $\omega_2 := \omega'\omega_1$ and $\omega'' := \omega_1^{-1}\omega$. Since $l(x)\omega_2$ contains the completely connected point B, we deduce from (16) that there exists a line x' in S such that $l(x)\omega_2 = l(x')$. Then we conclude $l\omega = l(x)\omega'\omega = l(x)\omega'\omega_1\omega_1^{-1}\omega = l(x)\omega_2\omega'' = l(x')\omega''$, and $l\omega$ is an ideal line by definition.

(18) $\Pi = (P, L, \in)$ *is a projective plane.*

(a) Let l, l' be any two distinct ideal lines in L. We choose points A, A' in l and l', respectively. By (15) there exists a product ω of elements in $K(O)$ such that $A\omega$, $A'\omega$ are proper points. From (17) we see that $l\omega$, $l'\omega$ are two ideal lines of the form $l(x)$, $l(x')$ with $x, x' \in S$ by (16). From $l \neq l'$ we deduce $x \neq x'$. Thus $S(xx')$ is the unique point in $l(x)$ and $l(x')$, and therefore $S(xx')\omega^{-1}$ is the unique point in l and l'.

(b) Let A, B be any two distinct points in P. Then we deduce from (15) that there is a product ω of elements in $K(O)$ such that $A\omega$ is a proper point. From $A \neq B$ we obtain $A\omega \neq B\omega$. Thus there exists exactly one element x in S such that $x \in A\omega$, $B\omega$. Then $A\omega$, $B\omega \in l(x)$ and therefore $A, B \in l(x)\omega^{-1}$. By (13) and (17) $l(x)\omega^{-1}$ is an ideal line. Hence A, B are contained in at least one ideal line. By the proof of (a) this ideal line is uniquely determined.

By (a) and (b) the axioms P1 and P2 are valid (compare the Appendix).

Since $E(G, S)$ contains a quadrilateral, the same is true for $(P, L, \in)$. Hence $(P, L, \in)$ is a projective plane.

(19) $\Pi = (P, L, \in)$ *is a pappian projective plane.*

The assumption of the theorem of Pappus-Pascal is as follows: Let a_i and b_k $(i, k = 1, 2, 3)$ be ideal lines in L, and A, B, C three points in P such that $A \in a_i$; $B \in b_k$; B, $C \notin a_i$. Moreover, let c_{ik} $(i, k = 1, 2, 3; i \neq k)$ be ideal lines such that $c_{ik} \in C$ and a_i, b_k, c_{ik} are concurrent. Finally, let $c_{12} = c_{21}$ and $c_{13} = c_{31}$. The assertion of the theorem is $c_{23} = c_{32}$. However, if A, B, C are not collinear, then we see from (15) that there is a product ω of elements in $K(O)$ such that $A\omega$, $B\omega$, $C\omega$ are proper points, and thus the assumptions of Theorem 5.12 are fulfilled. Hence our assertion follows immediately from Theorem 5.12. If A, B, C are collinear, a familiar indirect proof shows our assertion: Suppose that $c_{23} \neq c_{32}$; then the intersections A_{23}, A_{32} of the distinct lines a_2, b_3 and a_3, b_2, respectively, are distinct. Let c' be the joining line of A_{23} and A_{32} and let C' be a point incident with c' and $c_{12} = c_{21}$. Then $C' \neq C$ and A, B, C' are not collinear. For the new configuration with C' instead of C we have $c_{12}' = c_{12} = c_{21} = c_{21}'$ and $c_{23}' = c_{32}'$ if we denote the elements of the new configuration by a dash. Thus Theorem 5.12 yields, since $A' = A$, $B' = B$, C' are not collinear, that $c_{13}' = c_{31}'$. But this implies $c_{13}' = c_{13} = c_{31} = c_{31}'$ and therefore $C' = C$, and A, B, C would be not collinear, in contradiction to our assumption.

This completes the proof that the theorem of Pappus-Pascal is valid for the projective plane $(P, L, \in)$. Hence (19) is true.

We denote the projective incidence structure belonging to the projective plane $(P, L \in)$ by Π as well. Obviously, up to isomorphisms we may regard $E(G, S)$ as a substructure of Π. We show

(20)   *Any reflection $\neq 1$ of* $E(G, S)$ *is the restriction of a projective reflection of* Π.

Let a be any element in S. Then a^* fixes all the points of $l(a)$. If $a^* \neq 1$, then a is not in the center of G, hence there exists a uniquely determined point, that contains the set $a^{\perp}$. We denote this point by S_a. We wish to prove that the map $X \mapsto X^a = Xa^*$ coincides with the point map given by a projective reflection with the center S_a and the axis $l(a)$.

As proved under (5), there exists a proper point A such that $a \notin A$ and $A \neq S_a$. Since Π is a pappian projective plane there exists a perspective collineation σ in $P(S_a, l(a))$ such that $A\sigma = Aa^* = A^a$, since A, Aa^*, S_a are collinear.

Now let X be any point in P. We wish to show $X\sigma = Xa^*$. Obviously, if $X = S_a$ or $a \in X$, then $X\sigma = Xa^*$. Thus we may assume that $a \notin X$ and $X \neq S_a$. If X is a completely connected point such that $(A, S_a) \notin X$, then we can construct $X\sigma$ as follows: If $b = (X, A)$, then $X\sigma = (X, S_a) \cdot (A\sigma, b \cdot a)$. Clearly, all the lines just considered are ideal lines of the form $l(x)$. Hence the same construction is true for Xa^*, and therefore we obtain $X\sigma = Xa^*$. If $(A, S_a) \in X$ and X is a completely connected

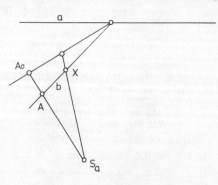

point $\neq S_a$ with $a \notin X$, we use a completely connected point not on (A, S_a) and a to deduce by a similar argument that $X\sigma = Xa^*$.

Finally, let X be an arbitrary point $\neq S_a$ not on a; then we choose any two proper points B, C such that X, B, C are noncollinear. This is possible by (4). Moreover, by Proposition 7.2 we can choose two proper points $D, E \neq B, C$ on the lines joining X with B and C. Then we have by our arguments just given above $B\sigma = Ba^*$, $C\sigma = Ca^*$, $D\sigma = Da^*$, $E\sigma = Ea^*$. This implies $(B, D)\sigma = (B, D)a^*$ and $(C, E)\sigma = (C, E)a^*$ and therefore $X\sigma = Xa^*$, since $(B, D) \neq (C, E)$ and $(B, D), (C, E) \in X$.

From $a^{*2} = 1$ we deduce $\sigma^2 = 1$, and therefore σ is a projective reflection of Π and a^* is the restriction of σ. This proves (20).

The pappian projective incidence structure Π belonging to the S-group plane $E(G, S)$ with completely connected points is called the *ideal incidence structure of* $E(G, S)$. Since the set of points in Π coincides with the set of points of $E(G, S)$, we have proved:

PROPOSITION 7.3. *Any S-group plane with completely connected points is properly embeddable in a pappian projective incidence structure that is its ideal incidence structure.*

By Proposition 7.3 up to isomorphisms we may regard an S-group plane with completely connected points as a substructure of the projective coordinate incidence structure $I(V)$ over a field K. Our Main Theorem 7.1 then follows immediately from Theorem 5.14. This completes the proof of the main theorem in this section.

5. REMARKS

Now we wish to add some remarks to elucidate the result obtained in Theorem 7.1.

1. The concept of an S-group plane with completely connected points is in one sense more general, in another sense less general, than the concept of a complete S-group plane. There exist wide classes of examples of S-group planes with completely connected points which are not complete S-group planes (compare Section 6). On the other hand, no Minkowskian plane is an S-group plane with completely connected points. Obviously, any elliptic plane and any Euclidean plane is an S-group plane with completely connected points. However, there are many substructures of elliptic or Euclidean planes which are not elliptic or Euclidean planes themselves, but which are S-group planes with completely connected points (compare Section 6).

2. Our main theorem may be regarded as a result concerning the plane

absolute geometry. In this direction it may be instructive to compare our result with other approaches to absolute geometry.

First, we must mention the famous book of D. Hilbert, "Grundlagen der Geometrie" [7]. If we call any plane satisfying Hilbert's axioms of plane absolute geometry a *Hilbert plane* then any Hilbert plane in a suitable interpretation is readily seen to be an S-group plane with completely connected points. However, given an arbitrary S-group plane, we cannot find a corresponding Hilbert plane in every case.

Secondly, we must refer to the important book of F. Bachmann, "Aufbau der Geometrie aus dem Spiegelungsbegriff" [2], which was one of the starting points for me. This book is based on a concept which we may call a *Bachmann group*. This is a pair (G, S) consisting of a group G and an invariant subset S of the set J of involutions in G such that S generates the group G and such that the following axioms are satisfied (in the formulation of these axioms $\alpha \mid \beta$ for $\alpha, \beta \in G$ stands for $\alpha\beta \in J$; capital letters denote elements in $S^2 \cap J$, and lowercase letters elements in S):

(B1)   *Given* A, B *then there exists an element* l *such that* A, B $\mid$ l.

(B2)   A, B $\mid$ l, m *implies* A = B *or* l = m.

(B3)   *If* a, b, c $\mid$ A, *then there exists an element* d *such that* abc = d.

(B4)   *If* a, b, c $\mid$ l, *then there exists an element* d *such that* abc = d.

(B5)   *There exist elements* l, m, n *such that* l $\mid$ m *and neither* l $\mid$ n *nor* m $\mid$ n.

Let (G, S) be any Bachmann group. Then every element in $S^2 \cap J$ may be called a *point* and every element in S a *line*. Moreover, if $A \mid b$, we say the point A is *incident* with the line b, and if $a \mid b$, we call the line a *orthogonal* to the line b. Thus we obtain a group plane corresponding to the Bachmann group, which we may call a *Bachmann plane*. As proved in reference 2, Section 4.5, any Bachmann group satisfies Axiom S, hence any Bachmann group is an S-group. The S-group plane $E(G, S)$ of a Bachmann group (G, S) differs slightly from the concept of a Bachmann plane. The set of lines in $E(G, S)$ coincides with the set of lines of the corresponding Bachmann plane, but the points in $E(G, S)$ are pencils of lines (compare Section 4.5 in reference 2) and not the elements in $S^2 \cap J$, as in the case of the Bachmann plane. However, we can prove that the Bachmann groups are exactly those S-groups with completely connected points such that any point $S(ab)$ in $E(G, S)$ with $ab = ba$ is completely connected.

Third, the basic paper [45] by E. Sperner must be cited, which further was a starting point for me. Sperner introduced our Axiom S and considered in reference 45 S-group planes with the additional property that any line is contained in at least three completely connected points.

Calling such an S-group plane a *Sperner plane*, we see that any Bachmann plane can be interpreted as a Sperner plane. Karzel showed in references 25 and 26 that the coordinate field of a Sperner plane can be a field of char 2. This implies that there exist Sperner planes that are not Bachmann planes (in a suitable interpretation). Comparing the concept of a Sperner plane with that of an S-group plane with completely connected points, we see in Section 6 that there exist classes of S-group planes $E(G, S)$ with completely connected points which are not Sperner planes. Moreover, we establish examples of S-group planes $E(G, S)$ with completely connected points for which the set S' of all lines contained in at least one completely connected point generates a proper subgroup of G.

The main result in reference 45 implies that for any Sperner plane the theorem of Desargues is valid in a form quite near to our formulation in Theorem 5.13. However, the proof given in reference 45 for the theorem of Desargues does not work in the general case of an arbitrary S-group plane with completely connected points. I do not know any other proof of Theorem 5.13 other than that given in Chapter 5. However, Theorem 5.13 is essential in the proof of our Main Theorem 7.1.

Finally, we remark that the idea to construct ideal lines by means of germs of perspective collineations is due to E. Ellers and E. Sperner in reference 19. In this paper Ellers and Sperner give a proof that every Sperner plane is embeddable in a pappian projective plane.

6. NORMALIZER PLANES

Our next goal is to present some special examples of S-group planes with completely connected points which are not Sperner planes and, consequently, which are not Bachmann planes. These examples are constructed as certain proper substructures of Euclidean planes. The construction is due to B. Bollow (compare reference 18).

We start with the field $\mathbf{Q}$ of rationals and consider a quadratic field $K := \mathbf{Q}(\sqrt{k})$ over $\mathbf{Q}$ such that k is an integer in $\mathbf{Q}$ with $k \equiv 1 \bmod 4$. Let M be the intersection of all subrings of K containing all the elements $2/(1 + x^2)$ for x in K. Then M is a ring in K. Investigations of F. Bachmann and B. Bollow show (compare references 17 and 18) that M consists of all the elements $a/b + (b/c)\sqrt{k}$ such that*

a, b, c, d are integers in $\mathbf{Q}$; $(a, b) = 1 = (c, d)$; b, d odd; if p is a prime divisor of b or d, then -1 or $-k$ is quadratic residue modulo p. $\left.\begin{array}{r}\\ \\ \\ \end{array}\right\}$ (1)

---

*In equation 1 (a, b) for any two integers in $\mathbf{Q}$ denotes the greatest common divisor of a and b.

Now let $V := K^3$ and let Q be the quadratic form over V such that*

$$Q(x, y, z) := x^2 + y^2 \qquad \forall (x, y, z) \in V$$

Then $I(V, Q)$ is a Euclidean coordinate plane. Any point A in $I(V, Q)$ is a set of all nonisotropic one-dimensional subspaces in V contained in a uniquely determined two-dimensional subspace. Identifying A with that two-dimensional subspace, we see that there exists a vector A' in V such that

$$X \in A \Leftrightarrow XA' = 0$$

if XA' denotes the familiar scalar product in V. We say that the vector A' is a representing vector of A.

According to Chapter 6, Section 5, the Euclidean coordinate plane is an affine incidence structure and a substructure of the projective incidence structure $I(V)$. We may assume that $\langle (0, 0, 1) \rangle$ is the line at infinity of $I(V, Q)$. Then any completely connected point A in $I(V, Q)$ can be represented by a vector of the form $(a, b, 1)$ with $a, b \in K$, since the completely connected points in $I(V, Q)$ are exactly the points in $I(V, Q)$ that are not parallel pencils and, consequently, that do not correspond to a point in $I(V)$ containing the line $\langle (0, 0, 1) \rangle$. Obviously, the vector $(a, b, 1)$ is uniquely determined by the point A, and therefore we may be allowed to denote that representing vector $(a, b, 1)$ by the same symbol A as the point.

By Theorem 6.15 the Euclidean coordinate plane $I(V, Q)$ can be regarded as a complete S-group plane $E(G, S)$. Then the lines in $I(V, Q)$ are the elements of S. Let P' be the set of points in $E(G, S)$ represented by all vectors $(a, b, 1)$ such that $a, b \in M$. Then we denote by S' the set of all lines s in $E(G, S)$ such that $P's^* = P'$. We wish to show that S' defines a substructure of $E(G, S)$ with the required properties. First, we prove

(a) $a \in B$ *and* $B \in P'$ *implies* $a \in S'$.

We must show that $A^a \in P'$ for any point A in P'. Let $(r, s, 1)$ be the representing vector of B. Then $a = \langle (u, v, -ur - vs) \rangle$ with $u, v \in K$ and u, v not both zero. Furthermore, let $(x, y, 1)$ be the representing vector of A and let $(x^*, y^*, 1)$ be the representing vector of A^a. Applying Chapter 1, Section 4, we deduce by an easy computation that

$$x^* = \frac{v^2 - u^2}{v^2 + u^2} x - \frac{2u^2}{v^2 + u^2} r + \frac{2uv}{v^2 + u^2} (s - y)$$

$$y^* = \frac{2v^2}{v^2 + u^2} s - \frac{v^2 - u^2}{v^2 + u^2} y + \frac{2uv}{v^2 + u^2} (r - x)$$

*As in Chapter 3, we write $Q(x, y, z)$ instead of $Q((x, y, z))$ for (x, y, z) in V.

Obviously, we have $2v^2/(v^2+u^2) \in M$ and $2u^2/(v^2+u^2) \in M$ and $(v^2-u^2)/(v^2+u^2) = 2v^2/(v^2+u^2)-1 \in M$. hence $x^*, y^* \in M$ and $A^a \in P'$.

In particular, we see from (a) that S' contains at least two lines. Thus S' defines a substructure of $E(G, S)$. Let G' be the group generated by S'. Then we show

(b) (G', S') *is an S-group.*

By definition, S' generates G' and consists of involutions. To prove Axiom S for (G', S'), let $a, b, x, y, z \in S'$ and $a \neq b$ and $abx, aby, abz \in J$. Then Axiom S for (G, S) yields $xyz \in S$. Since $P'x^* = P'$ and $P'y^* = P'$ and $P'z^* = P'$, we obtain $P'(xyz)^* = P'x^*y^*z^* = P'$ and therefore $xyz \in S'$, as required. This proves Axiom S and thus (b).

(c) $E(G', S')$ *is an S-group plane with completely connected points.*

First, we have to show that $E(G', S')$ contains a quadrilateral. To prove this it suffices by (a) to show that P' contains at least four points, no three of them concurrent. However, the vectors $(0, 0, 1)$, $(0, 1, 1)$, $(1, 0, 1)$, and $(1, 1, 1)$ represent such four points, since $0, 1 \in M$.

Next we must deduce the existence of at least one proper point. By (a) any point A in P' is a completely connected point in $E(G', S')$, since $A \subset S'$ and A is completely connected in $E(G, S)$. Let $S(ab) \in P'$; then $S(ab)$ is a proper point in $E(G', S')$: If $xy = yx$ for all x, y in $S(ab)$, then we conclude, applying Corollary 5.3 and Proposition 6.37 [looking at the fact that $S(ab)$ is completely connected in $E(G, S)$], that there is at least one line x in $S(ab)$ such that $S(ab) = x^\perp \cup \{x\} = S_x$. Here $x^\perp$ denotes the set of perpendiculars to x in $E(G, S)$. Proposition 6.37 then yields char $K = 2$ in contradiction to char $\mathbf{Q}(\sqrt{k}) = 0$. Thus $S(ab)$ is a proper point, and (c) is proved.

The substructure $E(G', S')$ in the Euclidean plane $E(G, S)$ is called the *normalizer plane* of the point set P'.

(d) $E(G', S')$ *is not a Sperner plane.*

We prove that there exist an infinite number of lines which are not contained in any completely connected point.

We choose any element r in K satisfying the following conditions:

$$\left.\begin{array}{l} r = \dfrac{a}{b} + \dfrac{c}{d}\sqrt{k}; \ a, b, c, d \quad \text{integers in} \quad \mathbf{Q} \\[2mm] (a, b) = 1 = (c, d); \ b, d \quad \text{not both even} \quad r \notin M; \ 2r \in M \end{array}\right\} \quad (2)$$

There exists an infinite number of such elements: For any odd integer $a \neq 0$ in $\mathbf{Q}$ the element $a/2 = a/2 + (0/1)\sqrt{k}$ in K fulfills our conditions.

We consider the line $s := \langle(-1, 0, r)\rangle$. Then $s \in S'$, since for any point A in P' with the representing vector $(u, v, 1)$ we obtain that As^* is represented by $(2r - u, v, 1)$, and therefore we have $As^* \in P'$.

We suppose that s is contained in a point B that is completely connected in $E(G', S')$. Let $(r, t, 1)$ be a vector representing B and let $a := \langle(u, v, -ur - vt)\rangle$ be an arbitrary line in B. By our assumption $a \in S'$. Since $0 \in M$, the vector $(0, 0, 1)$ represents a point O in P'. Hence $O^a \in P'$. A short computation yields that the vector

$$\left(\frac{2u^2}{u^2+v^2} r + \frac{2uv}{u^2+v^2} t, \; \frac{2uv}{u^2+v^2} r + \frac{2v^2}{u^2+v^2} t, \; 1 \right)$$

represents the point Oa^*. This implies

$$\frac{2u^2}{u^2+v^2} r + \frac{2uv}{u^2+v^2} t \in M \quad \text{and} \quad \frac{2uv}{u^2+v^2} r + \frac{2v^2}{u^2+v^2} t \in M \quad (3)$$

for all u, v in K such that u, v are not both zero. In particular, if we choose $(u, v) := (0, 1)$ or $(u, v) := (1, 1)$, we obtain

$$2t \in M \quad \text{and} \quad r + t \in M \quad (4)$$

Since $r \notin M$, we deduce from equation 4 that

$$t \notin M \quad (5)$$

We put

$$r = \frac{a}{b} + \frac{c}{d} \sqrt{k} \quad \text{and} \quad t = \frac{e}{f} + \frac{g}{h} \sqrt{k}$$

such that a, b, c, d, e, f, g, h are integers in $\mathbf{Q}$ and

$$(a, b) = (c, d) = (e, f) = (g, h) = 1 \quad (6)$$

Putting $(u, v) := (1, \sqrt{k})$, we obtain from equation 3

$$\frac{2(ah+kgb)}{(1+k)bh} + \frac{2(cf+de)}{(1+k)df} \sqrt{k} \in M \quad (7)$$

From equation 4 we deduce

$$\frac{af+be}{bf} + \frac{ch+dg}{dh} \sqrt{k} \in M \quad (8)$$

By equation 1 any power of 2 that divides one of the numerators in equation 7 or 8 also divides the corresponding denominator.

Now b, f, and h may each be even or odd. We show that each of these six possibilities leads to a contradiction:

(a) b even, f even, h even. Then equations 2 and 6 yield that d and e are odd. Since $4 \mid (1+k)df$,* we deduce from equation 7 that $2 \mid cf + de$, and therefore $2 \mid de$ and d or e is even, a contradiction.

(b) b even, f even, h odd. Then equation 6 yields that a is odd. Since $4 \mid (1+k)bh$, we have by equation 7 that $2 \mid ah + kgb$, and therefore $2 \mid ah$ and a or h is even, a contradiction.

(c) b even, f odd, h arbitrary. Then equation 6 implies that a is odd. We have $2 \mid bf$ and therefore $2 \mid af + be$ by applying equation 8. This yields $2 \mid af$, in contradiction to the fact that a and f are odd.

(d) b odd, f even, h even. From equation 6 we see that g is odd. Thus we have $4 \mid (1+k)bh$, and therefore by equation 7 $2 \mid ah + kgb$ and $2 \mid kgb$. This implies the contradiction that k or g or b must be even.

(e) b odd, f even, h odd. Then e is odd by equation 6. Since $2 \mid bf$, we conclude from equation 8 that $2 \mid af + be$ and thus $2 \mid be$, in contradiction to the fact that b and e are odd.

(f) b odd, f odd, h arbitrary. Then equation 5 yields that h is even, and from equation 6 we obtain that g is odd. Thus $4 \mid (1+k)bh$ and, successively, $2 \mid ah + kgb$ and $2 \mid kgb$, which contradicts the fact that k and g and b are odd.

Thus the line s is not contained in any completely connected point. Since there exist an infinite number of such lines, we have proved, in particular, statement (d).

---

* For any two integers x and y in $\mathbf{Q}$ the symbol $x \mid y$ means that x is a divisor of y.

8
FINITE S-GROUPS

An S-group (G, S) is called a *finite S-group* if the group G is finite.

Trivially, if the S-group (G, S) is finite, then S is finite. Conversely, by the Reduction Theorem 5.1 an S-group (G, S) with Δ-connected points is finite if S is finite. Thus an S-group (G, S) with Δ-connected points is finite if and only if S is finite.

An incidence structure (L, κ) is called a *finite incidence structure* if L is finite. Thus an S-group plane $E(G, S)$ with Δ-connected points is finite if and only if the S-group (G, S) is finite.

We recall some of the notations introduced in the preceding chapters.

Given any set M, then $|M|$ denotes the cardinality of M. Thus for a point X in an incidence structure the symbol $|X|$ denotes the cardinality of the set X of lines. Deviating from this slightly, we denote the cardinality of the set of points on the line x in an incidence structure by $|x|$.

If X is any point in an S-group plane, we denote by $U(X)$ the set of all the points which are not connected with X by a line.

1. ELEMENTARY REMARKS ON FINITE S-GROUPS

To prepare for the following proposition, which is very important for the study of finite S-group planes, we consider the following map:

Given any point A in an arbitrary S-group plane $E(G, S)$ and given any line a in A, we put

$$\mu_a : \begin{cases} A \to A \\ x \mapsto a^x \end{cases}$$

For this map we define

$$\ker \mu_a := \{x \mid x\mu_a = a\}$$

This set may be called the *kernel* of the map μ_a. Obviously, we have

$$\ker \mu_a = A \cap (a^\perp \cup \{a\})$$

$$\operatorname{Im} \mu_a = a^A$$

Since $x \in A \cap a^\perp$ is equivalent to $xab \in A \cap b^\perp$ for $a, b \in A$, and therefore $(A \cap a^\perp)ab = A \cap b^\perp$, we obtain

$$|\ker \mu_a| = |\ker \mu_b| \qquad \forall a, b \in A$$

We state the following:

LEMMA 8.1. *Let* A *be a point in a finite S-group plane and* $a \in A$. *Then*

$$|\ker \mu_a| \cdot |\operatorname{Im} \mu_a| = |A|$$

PROOF. If b is any line in $\operatorname{Im} \mu_a = a^A$, then the equation

$$a^x = b$$

has at least one solution x_0. Moreover, it has exactly $|\ker \mu_a|$ solutions, since for any solution x of the equation we have the following equivalent statements:

$$a^x = b \Leftrightarrow a^x = a^{x_0} \Leftrightarrow a^{a x_0 x} = a \Leftrightarrow a x_0 x \in \ker \mu_a \Leftrightarrow x \in (x_0 a) \ker \mu_a.$$

Thus our assertion follows from $|(x_0 a) \ker \mu_a| = |\ker \mu_a|$. Since every element x in A belongs to exactly one element b in $\operatorname{Im} \mu_a$ such that $a^x = b$, and since exactly $|\ker \mu_a|$ elements in A belong to the same element b in $\operatorname{Im} \mu_a$, we immediately obtain $|\ker \mu_a| \cdot |\operatorname{Im} \mu_a| = |A|$. ∎

PROPOSITION 8.2. *Let* A *be any point in a finite S-group plane. Then*

(a) $|A|$ *is even* $\Leftrightarrow$ *there exist* u, v *in* A *such that* $uv \in J$.

(b)   *If* $|A|$ *is odd, then* $a^A = A$ *for all* a *in* A.

(c)   *If* $|A|$ *is even and* $|A \cap b^\perp| \leq 1$ *for some* b *in* A *then* $|a^A| = \frac{1}{2}|A|$ *for all* a *in* A.

PROOF. For (a): We have the following sequence of equivalent statements: $|A|$ is even $\Leftrightarrow$ $|A^2| = |A|$ is even $\Leftrightarrow$ the group A^2 contains at least one involution $\Leftrightarrow$ there are u, v in A such that $uv \in J$.

For (b): If $|A|$ is odd, then $|A \cap a^\perp| = 0$ for all a in A by (a). Thus $|\ker \mu_a| = 1$, and therefore $|\operatorname{Im} \mu_a| = |A|$ and $a^A = \operatorname{Im} \mu_a = A$ by Lemma 8.1.

For (c): Let $|A|$ be even and $|A \cap b^\perp| \leq 1$ for some b in A. Then (a) implies that there are lines u, v in A such that $uv \in J$. Clearly, we have

$buv \in A \cap b^{\perp}$, and therefore $|A \cap b^{\perp}| = 1$ and $|\ker \mu_b| = 2$. By the remark made above we conclude $|\ker \mu_a| = 2$, and therefore $|a^A| = |\operatorname{Im} \mu_a| = \frac{1}{2}|A|$ for all a in A by Lemma 8.1. ∎

From Proposition 8.2(a) we immediately deduce:

COROLLARY 8.3. *Let* A *be any point in a finite S-group plane. Then*

$$|A| \text{ is even } \Leftrightarrow A \cap b^{\perp} \neq \varnothing \text{ for all } b \text{ in } A$$

LEMMA 8.4. *Let* E(G, S) *be any finite S-group plane. Then*

(a) $|X| \le |x|$ *for all points* X *and all lines* x *with* $x \notin X$.

(b) $|X| = |x|$ *for all completely connected points* X *and all lines* x *with* $x \notin X$.

(c) $|X| = |x| - 1$ *for all 2-Δ-connected points* X *and all lines* x *with* $x \notin X$.

(d) *Let* X *be a point. If* $|x| \le |X|$ *for all lines* x *with* $x \notin X$, *then* X *is completely connected.*

(e) *Let* X *be a point that is not completely connected. If* $|x| - 1 \le |X|$ *for all lines* x *with* $x \notin X$, *then* X *is 2-Δ-connected.*

PROOF. To prove (a), (b), and (c), let X be a point and let x denote a line such that $x \notin X$.

For (a): The intersections of the lines in X with x are mutually distinct points on x, hence $|X| \le |x|$.

For (b): If X is completely connected, then the points on x are all connected with X by mutually distinct lines. Hence $|x| \le |X|$, which by (a) implies that $|X| = |x|$.

For (c): If X is 2-Δ-connected, then Corollary 5.9 yields that there exists at least one point on x not connected with X, hence $|X| \le |x| - 1$. By definition of a 2-Δ-connected point, there is at most one point on x not connected with X, hence $|x| - 1 \le |X|$ and therefore $|X| = |x| - 1$.

For (d): Let X denote a point such that $|x| \le |X|$ for all lines x with $x \notin X$. Given any point Y, we show that $X \cap Y \neq \varnothing$: We choose any line y in Y. If $y \in X$, we are done. If $y \notin X$, then $|y| \le |X|$, and therefore the intersections of lines in X with y are all the points on y. In particular, this means that Y is the intersection of a line in X with y. This yields $X \cap Y \neq \varnothing$, as required. Hence X is completely connected.

For (e): Let X be any point such that $|x| - 1 \le |X|$ for all lines x with $x \notin X$. Then X is not connected with at most one point on all lines x with $x \notin X$. Hence X is completely connected or 2-Δ-connected. This yields (e).

Thus Lemma 8.4 is proved. ∎

In the following, two special methods of counting elements are used in finite S-group planes. The first can be presented as follows:

Let $E(G, S)$ be any finite S-group plane and let a be any line. We put $n := |a|$ and denote the points on a by $A_1, \ldots, A_n$. Then we have

$$|S| - 1 = \sum_{i=1}^{n} (|A_i| - 1) \qquad (*)$$

We speak in this case of "counting S by the intersections of lines with a."

Let $E(G, S)$ be any finite S-group plane, M any set of points in $E(G, S)$, and A a fixed point in $E(G, S)$ (A may or may not be in M). If $a_1, \ldots, a_n$ are all the lines in A, and if s_i denotes the number of points in M on a_i for $1 \le i \le n$, then

$$|M| = \begin{cases} |M \cap U(A)| + \sum_{i=1}^{n} s_i & \text{if} \quad A \notin M \\ |M \cap U(A)| + \sum_{i=1}^{n} (s_i - 1) + 1 & \text{if} \quad A \in M \end{cases} \qquad (**)$$

2. FINITE S-GROUP PLANES WITH COMPLETELY CONNECTED POINTS

We recall that an S-group plane with completely connected points is an S-group plane containing at least one quadrilateral and at least one completely connected point $S(ab)$ such that $ab \ne ba$.

As in Chapter 7, any completely connected point $S(uv)$ satisfying $uv \ne vu$ is called a proper point.

THEOREM 8.5. *Let* E *be any finite S-group plane* $\ne E(G_1, S_1)$. *Then the following are equivalent:*

(a) E *is an S-group plane with completely connected points.*

(b) E *is a Euclidean plane or a hyperbolic-metric plane.*

PROOF. (a) follows from (b): If E is a Euclidean plane, then it is an S-group plane with completely connected points, as we see from Theorem 6.15, Proposition 6.44, and the remarks in Chapter 2, Section 3. On the other hand, if E is a finite hyperbolic-metric plane $\ne E(G_1, S_1)$, then it is up to isomorphism a hyperbolic-metric coordinate plane $I(V, Q)$ over a finite field $K \ne GF(2)$. But then Lemma 2.5 guarantees the existence of a quadrilateral in $I(V, Q)$. Moreover, since K is finite, it can easily be shown that there exists a two-dimensional subspace T of the metric vector space (V, Q) which contains no isotropic vector $\ne O$, and which satisfies $g_Q(X, Y) \neq 0$ for at least two vectors X, Y in T. Hence there exists a proper point in $I(V, Q)$ (compare Chapter 6, Section 10), and so E is an S-group plane with completely connected points.

Next we prove that (a) implies (b). This proof is more complicated.

Let $E(G, S)$ be any S-group plane with completely connected points. Recall the following statements proved in Chapter 7, Section 2: Let A denote a proper point in $E(G, S)$; then

(1) A *does not contain three distinct lines* a, b, c *satisfying* $a, b \in c^{\perp}$.

(2) $A^x = A^y$ *and* $x, y \notin A$ *imply* $xy = yx$ *and, in particular,* $x = y$ *if there is no line* c *such that* $A \subset c^{\perp}$.

(3) $A \subset x^{\perp}$ *and* $A \subset y^{\perp}$ *imply* $x = y$.

(4) A^S *contains at least three noncollinear points.*

Since the proof of these statements in Chapter 7, Section 2, placed no restrictions on the cardinality of $E(G, S)$, we may apply them to our proof here.

We continue the proof. Fix a proper point A and two lines a, b in A such that $ab \neq ba$. These specific elements are referred to in the following whenever the symbols A, a, b occur. We first show

(5) $A^G = A^S$.

Since $S \subset S^3$, the Reduction Theorem 5.1 yields $G = S^2 \cup S^3$. From $A^{S^2} = A^{aS^2}$ and $aS^2 \subset S^3$ we see that $A^{S^2} \subset A^{S^3}$. Thus it suffices to show $A^{S^3} \subset A^S$. Let α be any element in S^3. Then (1) in the proof of Theorem 5.1 implies that there exist lines c, d in A and a line e such that $\alpha = cde$. Obviously, $A^{\alpha} = A^e$ and therefore $A^{S^3} \subset A^S$.

(6)   *For any line* x *we have* $|x| = |A|$.

Let x be any line. By (4), (5) there exists a point B in A^G such that $x \notin B$. Obviously, B is completely connected and therefore $|x| = |B| = |A|$ by Lemma 8.4(b).

In the following we sometimes write n instead of $|A|$. Clearly $n \geq 3$, since a, b, aba are three mutually distinct lines in A.

Now, we separate the proof into two cases. The first reads:

CASE A.   *There is no line* c *such that* $A \subset c^{\perp}$.

Obviously, Case A implies:

$$A^x = A \Leftrightarrow x \in A$$

Moreover, we show

(7)   *In Case A we have* $|A^G| = |S| - |A| + 1$.

From (5) we obtain $|A^G| = |A^S|$. By (2) we have $A \neq A^x \neq A^y$ for all $x \neq y$ in $S \backslash A$. Hence we deduce $|A^S| = |\{A\} \cup A^{S \backslash A}| = 1 + |A^{S \backslash A}| = 1 + |S \backslash A| = 1 + |S| - |A|$. This yields (7).

(8) *In Case A any line through* A *is contained in exactly* $|A| - 1$ *points of* A^G *and in exactly one 2-Δ-connected point. Consequently, all points are completely connected or 2-Δ-connected.*

We denote the maximal number of points in A^G on a line through A by r. Without loss of generality we may assume that a is contained in r points of A^G, as otherwise we interchange a, b with lines a', b' in A such that $ab = a'b'$ and then a' is contained in exactly r points in A^G.

Obviously, for any point X we have

$$|X| \geq r \tag{1}$$

since for $A \neq X$ the r points in A^G on a (if $a \notin X$) or on a^b (if $a \in X$) are connected with X by mutually distinct lines. If $X = A$, then equation 1 holds obviously. Denoting the points on a not in A^G by $B_1, \ldots, B_{n-r}$ and counting S by the intersections of lines with the line a, we obtain from $(*)$ and equation 1:

$$|S| - 1 = r(n-1) + \sum_{i=1}^{n-r} (|B_i| - 1) \geq n(r-1) + r(n-r) \tag{2}$$

Counting A^G from the point A, we see from $(*)$

$$|A^G| \leq (r-1)n + 1 \tag{3}$$

From (7) we see by combining equations 2 and 3 that

$$(r-1)n + 1 \geq |A^G| = |S| - |A| + 1 \geq n(r-1) + r(n-r) - n + 2$$
$$= n(r-1) + (n - (r+1))(r-1) + 1$$

This yields $r = 1$ or $r + 1 \geq n$. The first case is impossible: If $r = 1$, then $A^G = \{A\}$ and therefore $1 = |A^G| = |S| - |A| + 1$ and $S = A$, which contradicts the existence of a quadrilateral. Thus by applying equation 1, we see that $n - 1 \leq r \leq n$. If $r = n$, then equation 1 and (6), in combination with Lemma 8.4(d), yield that any point is completely connected. Thus $E(G, S)$ is an elliptic plane, and therefore there exists a polar triangle or a central line (compare Proposition 6.43). Familiar arguments and (4) then yield the existence of a line c such that $A \subset c^\perp$, a contradiction. Hence $r = n - 1$.

From (7) and equations 1, 2, and 3 we obtain

$$(n-1)^2 \leq (n-1)^2 + |B_1| - (n-1) = |S| - |A| + 1 = |A^G| \leq (n-1)^2$$

This implies $|B_1| = n - 1$ and $|S| = n(n-1)$ and $|A^G| = (n-1)^2$. If there existed any line x through A which was contained in less than $n - 1$ points of A^G, then $|A^G| < (n-1)^2$, by $(**)$. Hence any line x through A is contained in exactly $n - 1$ points of A^G. Counting S by the intersection of lines with x, we see from an equation analogous to that of equation 2 that the unique point on x, which is not contained in A^G, lies on $n-1$ lines, and therefore is an 2-Δ-connected point by Lemma 8.4(e) and (6). Thus any line through A is contained in exactly $n-1$ completely connected points (the points in A^G on x) and in exactly one 2-Δ-connected point. Thus (8) is proved.

The proof of (8) applied in Case A immediately yields:

$$|S| = |A|(|A|-1) \quad \text{and} \quad |A^G| = (|A|-1)^2 \tag{4}$$

(9) *In Case A the S-group plane* $E(G, S)$ *is a Euclidean plane.*

From (8) we deduce that the finite S-group plane $E(G, S)$ with completely connected points is a complete S-group plane in Case A. Moreover, it is a Euclidean plane, since by (8) Axiom-Δ1 is valid. Axiom EM is fulfilled, since there exists at least one 2-Δ-connected point B by (8), and therefore we can find a point C not connected with B by a line. That point C must be 2-Δ-connected by (8). Finally, Axiom$\neg$H is true, since any two distinct 2-Δ-connected points B, C cannot be connected by a line: If $y \in A \cap C$, then $y \notin B$ by (8), and the lines joining B and the $n-1$ points in A^G on y are all the lines in B and therefore $B \cap C = \varnothing$.

Thus the assertion of Theorem 8.5 is valid in Case A.

Next we deal with

CASE B. *There is a line* c *such that* $A \subset c^\perp$.

By (3) the line c is uniquely determined. In the following, whenever the symbol c occurs we are referring to this uniquely determined line. We prove

(10) *If* $|A|$ *is odd in Case B then* c *is a central line.*

Let $|A|$ be odd and let x denote any line. If $x \in A \cup \{c\}$, then clearly $xc = cx$. If $x \notin A \cup \{c\}$, then there exists a line z in $A \cap S(cx)$, and by Proposition 5.11 a line y in $A \cap x^\perp$. If $y \neq z$, then $y \notin S(cx)$ and $S(cx) \subset y^\perp$. In particular, we have $yz \in J$ and $A = S(yz)$, and therefore $|A|$ is even by Proposition 8.2(a). This contradicts our assumption, and thus we obtain $y = z$. But then $cz = zc$ and $xz = zx$ and $cxz = zxc$ and therefore $xc = cx$. Hence c is a central line.

From (10) and Lemma 6.35 we immediately deduce

(11) *If* $|A|$ *is odd in Case B, then* $S(cx) = x^\perp \cup \{x\}$ *for all* $x \neq c$.

Next we show

(12) *If* $|A|$ *is even in Case B, then* $x \in S^2$. *Also* $x^\perp$ *is a point and* $|x^\perp|$ *is even for all* x.

Let $|A|$ be even. Then an application of Proposition 8.2(a) ensures the existence of a line d such that $A = S(ad)$ and $ad \in J$. Obviously, $adc = 1$.

Let x be a line. We claim that $x \in S^2$: This is clear if $x = c$. If $x \neq c$, then there exists a line y in $A \cap S(cx)$. Since ady, $ycx \in S$, we obtain $x = adcx = (ady)(ycx) \in S^2$. This implies $x^\perp = S(x)$, and $|x^\perp|$ is even by Proposition 8.2(a).

In the following, given $x \in S$, we put

$$S_x := \begin{cases} S(xc) & \text{if } |A| \text{ is odd and } x \neq c \\ x^\perp & \text{if } |A| \text{ is even} \end{cases}$$

From Proposition 8.2(a) and by (12) we see that $|S_x|$ is even whenever S_x is defined.

(13) $x \in A$ *implies* $A \neq S_x$ *in Case B.*

If $x \in A$ and $A = S_x$, then $x \in S_x$. Therefore $|A|$ is odd and also even, since $A = S_x$, a contradiction. Hence $A \neq S_x$.

(14) $x \in A$ *and* $x \in B \in A^G$ *in Case B imply* $B \in A^{S_x}$.

Let $x \in B \in A^G$. Then we deduce from (5) that there exists a line y such that $B = A^y$. If $y \in A$, then $B = A$. We choose a line u in $A \cap S_x$ and obtain $B = A = A^u \in A^{S_x}$. If $y \notin A$, then $x \in A$, A^y implies $y \in x^\perp \subset S_x$. Thus in either case we have $B \in A^{S_x}$.

(15) *In Case B we have* $|A^{S_x}| = \frac{1}{2}|S_x|$ *for* $x \in A$.

Let B be any point in A^{S_x}. Then our assertion follows from the fact that the equation

$$A^y = B \tag{5}$$

has exactly two solutions in S_x. To prove this let u be the line in $A \cap S_x$ which is uniquely determined by (13).

If y_0 is a solution of equation 5 in S_x, then cuy_0 also is a solution in S_x, since $A^{cuy_0} = A^{y_0} = B$. Obviously, $y_0 \neq cuy_0$. Thus there exist at least two distinct solutions of our equation.

Let y be an arbitrary solution of equation 5; then $A^{uy_0y} = A$. This implies $uy_0y \in A$ or $A \subset (uy_0y)^\perp$. In the first case we conclude from $u, uy_0y \in A$, S_x, and $A \neq S_x$ [by (13)] that $u = uy_0y$ and therefore $y = y_0$. In the second case we obtain $uy_0y = c$ by (3), and therefore $y = cuy_0$. Thus the only solutions of equation 5 are y_0 and cuy_0. This proves (15).

(16) *If* $|A|$ *is odd in Case B, then* $a^G = S\backslash\{c\}$.

Let x be any line $\neq c$. We may assume $x \neq a$, since $a \in a^G$. If $|S(ax)|$ is odd, then $x \in a^S$ by Proposition 8.2(b). If $|S(ax)|$ is even, then $c \in S(ax)$, since c is a central line by (10). There exists a line d such that $xad \notin S$. Then $c \notin S(ad)$, $S(dx)$, and therefore $|S(ad)|$ and $|S(dx)|$ are odd. Applying Proposition 8.2(b), we conclude $x \in d^S$ and $d \in a^S$. Therefore $x \in a^{S^2}$, as required.

(17) *If* $|A|$ *is odd in Case B, then* $|B| = |A| - 1$ *for all points* B *on* c *and* $|C| = |A| - 2$ *for all points* C *not in* A^G *and not on* c.

Let r denote the number of points in A^G on the line a. Then we deduce from (16) that every line $\neq c$ is contained in r points of A^G. Obviously, no point in A^G lies on c, since $c \notin A$ and c is a central line. Counting A^G from A, from B, and from C (for a point B on c and a point C not in A^G and not on c), we obtain from $(**)$

$$|A^G| = |A| \cdot (r-1) + 1 = (|B| - 1) \cdot r = |C| \cdot r \qquad (6)$$

From (14) and (15) we see that $r = \frac{1}{2}|S_a| = \frac{1}{2}|S_x|$ for all $x \neq c$ and therefore $r = \frac{1}{2}|B|$, since $B = S_x$ for all $x \neq c$ in B by (11). If equation 6 is used, an easy computation now yields that $|B| = |A| - 1$ and $|C| = |A| - 2$.

Let $|A|$ be even. Then $|A \cap a^{\perp}| = 1$ by (12). Thus we conclude from Proposition 8.2(c) that there exists a line d in A such that

$$A = a^A \cup d^A \qquad (7)$$

For the following, let d always denote a line for which equation 7 holds. Without loss of generality we may assume that $|a^{\perp}| \geq |d^{\perp}|$.

(18) *If* $|A|$ *is even in Case B then* c *is contained in at least one point of* A^G.

Suppose the contrary. By (12) $a^{\perp}$ is a point B on c and $|B|$ is even. Since $|B \cap c^{\perp}| = 1$, we may use Proposition 8.2(c) to deduce the existence of a line e such that $B = c^B \cup e^B$. Then e is contained in at least one point of A^G, since this is true for the line f joining B with A, and therefore we have $f \in e^B$ (as c is not contained in any point in A^G and thus $f \notin c^B$). Counting A^G from A and from B, we apply $(**)$ and (15) to obtain

$$\tfrac{1}{2}|A| \left(\tfrac{1}{2}|a^{\perp}| + \tfrac{1}{2}|d^{\perp}| - 2 \right) + 1 = |A^G| = \tfrac{1}{2}|B| \cdot \tfrac{1}{2} \cdot |e^{\perp}|$$

If $|d^{\perp}| \geq 3$, then $|d^{\perp}| \geq 4$, since $|d^{\perp}|$ is even by (12). Furthermore, we obtain $|A||a^{\perp}| + 4 \leq |A|(|a^{\perp}| + |d^{\perp}| - 4) + 4 = 4|A^G| = |a^{\perp}||e^{\perp}| \leq |a^{\perp}||A|$ and therefore $4 \leq 0$, a contradiction. If $|d^{\perp}| < 3$, then $|d^{\perp}| = 2$. However, this yields $|a^{\perp}| = 2$, as otherwise there would exist two points in A^G on a (or on a^b) which are connected with $d^{\perp}$ by distinct lines $\neq c$ and thus we would obtain $|d^{\perp}| \geq 3$. From $|a^{\perp}| = 2$ we see that $|X| = 2$ for all X on c and A is the only point not on c, in contradiction to the existence of a quadrilateral. Thus our assumption is false, and (18) is proved.

If $|A|$ is even, we see from (18) that in Case B we may assume without loss of generality that $a^{\perp}$ belongs to A^G. Consequently, we have $|a^{\perp}| = |A|$. Moreover, we have $d^{\perp} \notin A^G$, as otherwise all points on c would be points of A^G, in contradiction to (14) and (15).

(19) *If* $|A|$ *is even in Case B, we have* $|d^{\perp}| = |A| - 2$ *and* $|A^G| = \frac{1}{2}(|A| - 1)(|A| - 2)$.

Put $D := d^\perp$. By Proposition 8.2(c) there exists a line e such that $D = c^D \cup e^D$. Applying $(**)$ and (15), we count A^G from A and from D and obtain

$$|A| (|a^\perp| + |d^\perp| - 4) + 4 = 4 |A^G| = |D| (|c^\perp| + |e^\perp|)$$

This yields $(|A| - 2)^2 = |D| \, |e^\perp|$, since $|a^\perp| = |A|$, $D = d^\perp$, and $A = c^\perp$. Since $2 \le |D| \le |A|$ and $2 \le |e^\perp| \le |D|$, we conclude $|D| = |A| - 2$, $|e^\perp| = |A| - 2$, and $|A^G| = \frac{1}{2}(|A| - 1)(|A| - 2)$.

(20)   Let $|A|$ be even in Case B and let X be a point that is not completely connected. Then $|X| = |A| - 1$ if $|X|$ is odd, and $|X| = |A| - 2$ if $|X|$ is even.

Under the assumption of (20) we denote the (uniquely determined) line in $A \cap X$ by x.

If $|X|$ is odd, then $|x^\perp| < |A|$, as otherwise $X \cap x^\perp \ne \varnothing$ (since $|x| = |A|$), which would imply that $|X|$ is even by Proposition 8.2(a). Thus $x \notin a^A$, and therefore $x \in d^A$ and $|x^\perp| = |A| - 2$ by (19). Any line in X is contained in the same number of points in A^G by Proposition 8.2(b). Thus counting A^G from X, we obtain by applying (19): $\frac{1}{2}(|A| - 1)(|A| - 2) = |A^G| = |X| \cdot \frac{1}{2}(|A| - 2)$ and therefore $|X| = |A| - 1$, as required.

If $|X|$ is even and if $x^\perp \in A^G$, then we interchange the roles of A and $x^\perp$ in the proof of (19) to obtain $|X| = |A| - 2$, since $X \notin A^G$. If $x^\perp \notin A^G$, then $x \in d^A$, and therefore $|x^\perp| = |A| - 2$ by (19). There exists a line y in X such that $X = x^X \cup y^X$ by Proposition 8.3(c). Counting A^G from X and applying (19), we obtain

$$|X| (|x^\perp| + |y^\perp|) = 4 |A^G| = 2(|A| - 1)(|A| - 2)$$

Since $|y^\perp| \le |A|$, we deduce from this equation that $(|A| - 1)(|A| - 2) \le |X| (|A| - 1)$ and therefore $|A| - 2 \le |X| \le |A|$. If $|X| = |A|$, then X would be completely connected [compare Lemma 8.4(d)], a contradiction. Thus we conclude that $|X| = |A| - 2$, since both $|X|$ and $|A|$ are even. This proves (20).

We conclude the proof of our theorem by showing

(21)   In Case B every point is Δ-connected.

Let X be any point. If $|X| = |A|$, then X is completely connected by (6) and Lemma 8.4(d).

If $|X| = |A| - 1$, then X is 2-Δ-connected by Lemma 8.4(e).

Thus we are left with the case that $|X| \ne |A|, |A| - 1$. From (17) and (20) we see that $|X| = |A| - 2$. We show that X is 1-Δ-connected:

There exists a line x such that $X \subset x^\perp$: If $|A|$ is odd, then $|X|$ is odd, and therefore $c \notin X$ by (17). Thus we have $X \subset c^\perp$ by (10). If $|A|$ is even, then we choose u, v in X such that $uv \in J$. Let s be a line in $X \cap A$; then $uv = (uvs)(sc)c \in S^3 \cap J$, since $c \in S^2$ and $A = S(c)$. This implies $uv \in S$ and $X = (uv)^{\perp}$, and we may put $x := uv$. Obviously, x is uniquely determined.

There are exactly two distinct points in $U(X)^*$ on x, say E and F, since $|X| = |A| - 2$ and $x \notin X$ and $|x| = |A|$ by (6). The points E and F are 2-Δ-connected, as we see from (17) and (20) ($|E|, |F|$ are odd if $|A|$ is even).

Now let U, V, W be any three mutually distinct points that are pairwise connected by a line. We have to show that X is connected with at least one of these three points.

If U, V, W are collinear, then our assertion follows from $|y| = |A|$ for any line y and from $|X| = |A| - 2$. Thus we may assume that U, V, W are noncollinear. We show

$$\text{If} \quad U = E \quad \text{and} \quad V = F, \quad \text{then} \quad X \cap W \neq \varnothing \tag{8}$$

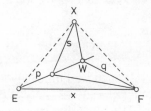

Let p, q denote the lines in $U \cap W$ and $V \cap W$, respectively. Then $W = S(pq)$. We consider the map

$$\mu : \begin{cases} X \to S \\ s \mapsto p^s \end{cases}$$

Since $p \neq x$, this map is injective. Moreover, we have $\text{Im } \mu \subset F$, since $E^s = F$ for all s in X [as otherwise we would obtain successively: $E^s = E$ and $F^s = F$; $E \subset s^{\perp}$ and $F \subset s^{\perp}$; s is a central line; $s = x$ by (10) and (12), $x \in X \subset x^{\perp}$, a contradiction]. Thus we have $|\text{Im } \mu| = |X| = |A| - 2$, $|F| = |A| - 1$, and $x \notin \text{Im } \mu$ (as $x^s = x \neq p$ for all s in X). Hence $q \in \text{Im } \mu$, and so we conclude $q = p^t$ for a suitable t in X. Therefore $t \in X \cap W$.

All other cases for U, V, W can be reduced to that shown in equation 8: If one of the points U, V, W is equal to E or to F, say $U = E$, then the 2-Δ-connected point F is connected with V or with W, say $F \cap V \neq \varnothing$. Equation 8 implies $X \cap V \neq \varnothing$ (also in the case $x \in V$), as required. If none of the points U, V, W is equal to E or to F, then the 2-Δ-connected point E is connected with at least two of the points U, V, W, say

_____
* We recall that $U(X)$ denotes the set of all points not connected with the point X.

$U \cap E \neq \emptyset \neq V \cap E$. Without loss of generality the 2-Δ-connected point F is connected with U. Hence $U \cap X \neq \emptyset$ by equation 8. This proves the assertion that X is 1-Δ-connected. Thus (21) is valid.

(22)   *In Case B the S-group plane* E(G, S) *is a hyperbolic-metric plane.*

By (21) the given finite S-group plane $E(G, S)$ with completely connected points is a complete S-group plane in Case B. Moreover, $E(G, S)$ is a hyperbolic-metric plane in Case B: (17), (19), and the proof of (21) show that there exists at least one 1-Δ-connected point. Thus Axiom Δ1 is valid. Further, the proof of (21) shows that there exist two distinct 2-Δ-connected points connected by a line. Hence Axiom H is fulfilled.

From (9) and (22) we deduce the assertion that (a) implies (b). This concludes the proof of Theorem 8.5. ∎

The following corollaries follow immediately from the proof of Theorem 8.5, in particular, from (9), equation 4, and (22), and from Chapter 6, Section 11.

COROLLARY 8.6.   *Let* E(G, S) *be a finite S-group plane with completely connected points and let* A *be a proper point in* E(G, S).

(a)   E(G, S) *is a Euclidean plane if and only if there exists no line* c *such that* $A \subset c^{\perp}$.

(b)   E(G, S) *is a hyperbolic-metric plane if and only if there exists a line* c *such that* $A \subset c^{\perp}$.

(c)   E(G, S) *is a complete S-group plane of char* $\neq 2$ *if and only if* $|A|$ *is even.*

(d)   E(G, S) *is a complete S-group plane of char* 2 *if and only if* $|A|$ *is odd.*

COROLLARY 8.7.   *Let* E(G, S) *be a finite S-group plane with completely connected points and let* A *be a proper point. Then*

(a) $|A| \geq 3$.

(b)   *If* E(G, S) *is a Euclidean plane, then* $|A^{G}| = (|A| - 1)^2$.

(c)   *If* E(G, S) *is a hyperbolic-metric plane, then* $|A^{G}| = \frac{1}{2}(|A| - 1)(|A| - 2)$.

3. FINITE S-GROUP PLANES WITH 1-Δ-CONNECTED POINTS

Our next goal is to determine all the finite S-group planes containing at least one 1-Δ-connected point. Obviously, such an S-group plane contains at least one quadrilateral, since there exist two points not connected by a line [compare Chapter 2, Section 2(1)].

THEOREM 8.8. *Let* E *be any finite S-group plane* $\neq E(G_0, S_0)$, $E(G_1, S_1)$. *Then the following are equivalent:*

(a) E *contains at least one* 1-Δ-*connected point.*

(b) E *is a hyperbolic-metric plane or a Minkowskian plane.*

PROOF. From the definition of hyperbolic-metric planes and Minkowskian planes we see that (b) implies (a).

Conversely, let $E(G, S)$ be any finite S-group plane containing a 1-Δ-connected point, say A. Again, whenever we use the symbol A in the following, we refer to this point.

We separate the proof that $E(G, S)$ is a hyperbolic-metric plane or a Minkowskian plane into several parts. The general assumptions made for the special part are not repeated in the statements (1), (2), ..., given in that part. The assumptions made successively read as follows:

CASE A. *There exists a line* a *such that* $A \subset a^{\perp}$.

CASE B. *There does not exist a line* a *such that* $A \subset a^{\perp}$.

$$(\alpha) \quad A^G \cap U(A) \neq \varnothing$$
$$(1) \quad |A| \text{ is even.}$$
$$(2) \quad |A| \text{ is odd.}$$
$$(\beta) \quad A^G \cap U(A) = \varnothing.$$

As we shall see, Case B(β) is not possible.

We begin by discussing *Case A*.

The symbol a in the following denotes a fixed line such that $A \subset a^{\perp}$. From Corollary 5.9 we deduce that there exists a point B in $U(A)$ on a. For brevity's sake we write $A \perp B$ if $x \perp y$ for all $x \in A$ and for all $y \in B$. If $A \perp B$ does not hold, then we write $A \not\perp B$. First, we prove

(1) $A \subset a^{\perp}$ *and* $A \subset b^{\perp}$ *and* $b \in B$ *imply that* $a = b$.

Suppose that $a \neq b$. Then $B = S(ab)$ and $A \perp B$. We show:

There exists no central line: If z is a central line and if a', a'' are two distinct lines in A and b', b'' are two distinct lines in B, then a', $a'' \neq b'$, b'', since $A \cap B = \varnothing$. From (1) in the proof of Theorem 5.4 we see that

$$z \in S(a'b'), S(a'b''), S(a''b'), S(a''b'')$$

Obviously, we have $S(a'b'), S(a''b') \neq S(a'b''), S(a''b'')$, and a', $z \in S(a'b')$, $S(a'b'')$. Also a'', $z \in S(a''b'), S(a''b'')$. This yields $a' = z = a''$, a contradiction to the choice of a' and a''.

If $b' \in B$, then B is the only point in $U(A)$ on b': Suppose C is another point in $U(A)$ on b'. Then $C^x = C$ for all x in A, since $(b')^x = b'$ and $B^x = B$. This implies $B \cup C \subset x^{\perp}$, and therefore x is a central line for all x in A by Theorem 5.4. This contradicts what we just proved above.

Since A is 1-Δ-connected, there exists a line c and two distinct points C, D in $U(A)$ on c. By the argument just given above, we have $c \notin B$ and $A \cap S(ac) \neq \varnothing$. Let $d \in A \cap S(ac)$ and $q := cad$. Then $q \notin A, B$, hence the mapping $\sigma := \sigma_{qAB}$ is defined. From $A \cap S(qx) \neq \varnothing$ for all x in B we see that $|\mathrm{dom}\,\sigma| \geq |B| \geq 2$ and therefore A^{σ} does exist. Without loss of generality we may assume $A^{\sigma} = C$, since $c = d\sigma \in A^{\sigma}$. We have $q(x\sigma) = xx_B \in J$ for all x in A and therefore $A^{\sigma} \subset q^{\perp}$. Hence $C^q = C$ and, moreover, $D^q = D$, since $c \in D^q$ and $D^q = D^{cq} = D^{ad}$ and $A \cap D^{ad} = A^{ad} \cap D^{ad} = \varnothing$. This implies $C \cup D \subset q^{\perp}$, and q would be a central line by Theorem 5.4. This contradicts the statement made at the beginning of that proof. Thus $a \neq b$ is impossible and so (1) is proved.

(2) $A^{\alpha} = A^{\beta}$ for $\alpha, \beta \in B^2$ implies $\alpha = \beta$.

Choose α, β in B^2 and $A^{\alpha} = A^{\beta}$. Then put $\gamma := \alpha\beta^{-1}$ and obtain $A^{\alpha\gamma} = A$ and $a\gamma \in B$. Since $A \cap B = \varnothing$, we have $a\gamma \notin A$, and therefore $A \subset (a\gamma)^{\perp}$ and $a\gamma = a$ by (1). This yields $\gamma = 1$ and $\alpha = \beta$.

(3) $x, y \in A$ implies $B^x \neq B$ and $B^x = B^y$.

Suppose the contrary. Then there exist lines x, y in A such that $B^x = B$ or $B^x \neq B^y$. In the second case we see that $a \in B, B^x, B^y$. Also A is Δ-connected, hence $B = B^x$ or $B = B^y$. Thus it is sufficient to treat the first case $B^x = B$. Then $B \subset x^{\perp}$. Let z be any line $\neq x$ in A. If $B^z = B$, then $B \subset z^{\perp}$ and $A = S(xz) \subset w^{\perp}$ for all w in B, in contradiction to (1). Hence $B^z \neq B$. However, $B^x = B$ and $a \in B, B^z, B^{zx}$ and $B, B^z, B^{zx} \in U(A)$ imply $B^{zx} = B^z$. This yields $B \cup B^z \subset x^{\perp}$, and x is a central line by Theorem 5.4. From (1) in the proof of Theorem 5.4 we deduce that $xaz \in S$, and therefore $x, z \in A$, $S(xa)$ and $A \neq S(xa)$. We conclude $x = z$, a contradiction to the choice of z. Thus our assumption is false, and (3) is proved.

We now denote by C the point B^x. In (3) we saw that this point is uniquely determined.

(4)   *We have $|x| \leq |A| + 2$ for all lines* x.

This is clear if $x \notin A$. Now let $x \in A$ and b be any line $\neq a$ in B. Then $x^b \notin A$, as otherwise $x \in A$, A^b and $b \notin A$ and therefore $x \perp b$. This yields $B^x = S(ab)^x = S(a^x b^x) = S(ab) = B$ and $x \in A$, in contradiction to (3). Hence $|x| = |x^b| \leq |A| + 2$.

(5)   *We have $|A^{B^2}| = |B|$ and $|B| = |A| + 1$.*

The first statement immediately follows from (2).

To prove the second statement we choose any line $b \neq a$ in B and consider the map

$$\mu : \begin{cases} A^2 \to B \\ \alpha \mapsto b^\alpha \end{cases}$$

Since $B^{A^2} = B$ by (3), we see that $b^\alpha \in B$ for $\alpha \in A^2$. The map μ is injective: $b^\alpha = b^\beta$ implies $b^\gamma = b$ for $\gamma := \alpha\beta^{-1} \in A^2$. If $\gamma \neq 1$, then $A = S(\gamma)$ and $A^b = S(\gamma^b) = S(\gamma) = A$, and therefore $A \subset b^\perp$ for $b \in B$. This implies $a = b$, in contradiction to the choice of b. Hence $\gamma = 1$ and $\alpha = \beta$.

From the injectivity of μ we obtain $|B| \geq |\operatorname{Im} \mu| = |A^2| = |A|$. Since $a \notin \operatorname{Im} \mu$ (as otherwise $a = b^\alpha$ and $b = a^{\alpha^{-1}} = a$ for a suitable α in A^2), we have $|B| \geq |A| + 1$. If $|B| > |A| + 1$, then Lemma 8.4(d) and (4) show that B would be completely connected, in contradiction to $A \cap B = \varnothing$. Hence $|B| = |A| + 1$.

From Lemma 8.4(e) and (4) and (5) we see that B is a 2-Δ-connected point. Next we show

(6) $U(B) = A^{B^2}$.

Obviously, we have $A^{B^2} \subset U(B)$, since $A \in U(B)$. To prove $U(B) \subset A^{B^2}$ we count S by enumerating all the lines in points in $U(B)$ and in B. By Corollary 5.9 any line not in B is contained in exactly one point in $U(B)$. Let p denote the number of lines in points in $U(B)$ which are not in A^{B^2}. Then observing (5), we have

$$|S| = (|A| + 1)|A| + |A| + 1 + p = (|A| + 1)^2 + p \tag{1}$$

On the other hand, we can count S by the intersection of lines with a. If X is a point on a such that $|X| \geq |A| + 2$, then X is completely connected by Lemma 8.4(d) and (4). Thus X is connected with all the points in $U(B)$. Since any line $\neq a$ in X is contained in exactly one point in $U(B)$, we have $|U(B)| = |A| + 1$, and therefore by (5) $U(B) = A^{B^2}$. This implies $p = 0$. If there exists no point X on a for which $|X| \geq |A| + 2$, then we have $|X| \leq |A| + 1$ for all X on a and therefore, observing that $|B| = |C| = |A| + 1$, we obtain

$$|S| \leq (|A| + 2)|A| + 1 = (|A| + 1)^2 \tag{2}$$

From equations 1 and 2 we deduce that $p = 0$ and therefore $U(B) \subset A^{B^2}$ in any case. This yields $U(B) = A^{B^2}$.

(7)   *We have* $\{A\} \cup U(A) = \{B\} \cup U(B) \cup \{C\} \cup U(C)$.

We put $L(X) := \{X\} \cup U(X)$ for all points X. Then the assertion is $L(A) = L(B) \cup L(C)$.

Let $X \in L(A)$. If $X = A$, then $X \in L(B) \cup L(C)$. If $X \neq A$ and $X \notin L(B) \cup L(C)$, then X, B, C are three mutually distinct points in $U(A)$ which are pairwise connected by a line, in contradiction to the fact that A is Δ-connected. Hence $X \in L(B) \cup L(C)$ and $L(A) \subset L(B) \cup L(C)$.

Conversely, let $X \in L(B) \cup L(C)$. If $X = A$, then $X \in L(A)$. If $X \neq A$, then $X \in U(A)$. To see this, observe that $A \in L(B)$ and $A \in L(C)$. Also any two distinct points in $L(B)$ and any two distinct points in $L(C)$ are not connected by a line, since B and therefore $C \in B^G$ are 2-Δ-connected. Hence $L(B) \cup L(C) \subset L(A)$, and (7) is proved.

(8)   *In Case A the S-group plane* $E(G, S)$ *is a hyperbolic-metric plane.*

First, note that $E(G, S)$ is a complete S-group plane: Let X be any point in $E(G, S)$. If $X \in U(A)$, then, since A, B, C are Δ-connected, we invoke (6) and (7) to conclude that X is Δ-connected. This also implies that X is Δ-connected if $X \in U(Y)$ and $Y \in A^G$. Thus we may assume $X \notin U(Y)$ for all Y in A^{B^2}. The lines joining X with the points in A^{B^2} are mutually distinct, since any two distinct points in A^{B^2} are not connected by a line. Hence we deduce from (5) that $|X| \geq |A^{B^2}| = |A| + 1$, and X therefore is Δ-connected by (4) and Lemma 8.4(e).

Next we show that $E(G, S)$ is a hyperbolic-metric plane: By our general assumption Axiom Δ1 holds. Moreover, the points B and C are two distinct 2-Δ-connected points connected by the line a, and therefore Axiom H is valid. This proves (8).

Next, we wish to treat *Case B*.

Recall that in this case there exists no line x for which $A \subset x^{\perp}$. Then we have for all x

$$A^x = A \quad \text{implies} \quad x \in A$$

In addition we have

$$x \neq y \quad \text{and} \quad A \cap S(xy) \neq \varnothing \quad \text{and} \quad x \notin A \quad \text{imply} \quad A^x \neq A^y.$$

If $A^x = A^y$ and $z \in A \cap S(xy)$, then $A^{zxy} = A$ and $zxy \in S$. This yields z, $zxy \in A$, $S(xy)$ and $A \neq S(xy)$ (as $x \notin A$) and therefore $z = zxy$ and $x = y$, a contradiction to one of our assumptions.

In the following the symbol a denotes a fixed line that is contained in two distinct points B, C in $U(A)$. The symbols B and C, respectively, are reserved for these two points in $U(A)$ on a.

Obviously, we have $a \notin A$ and

$$|a| = |A| + 2$$

We prove

(9)   *There exists no central line.*

Suppose to the contrary there exists a central line z. By the assumption of Case B that line must be a line of A.

By Proposition 8.2(a) $|X|$ is even for any point on z. In particular, $|A|$ is even. On the other hand, $|X|$ is odd for any point not on z, as otherwise we would deduce from Proposition 8.2(a) that $X = S(xy)$ and $xy \in J$. However, this would yield $xyz = 1$, in contradiction to Corollary 4.6.

Since $z \notin B, C$, the number $|B|$ is odd, and therefore we have $|B| \geq 3$. We can choose two distinct lines b, $d \neq a$ in B and a line $c \neq a$ in C. Since A is connected with $S(bc)$ and with $S(dc)$, we have $|A| \geq 3$ or, without loss of generality, $|S(dz)| \geq 3$ [if $z \in S(cd)$]. We put $D := A$ or $D := S(dz)$, respectively, and obtain $|D| \geq 3$ and $z \in D$. Now let x, y be two lines $\neq z$ in D. Then $x \neq a$ and $z \notin S(ax)$. Hence $|S(ax)|$ is odd, and therefore $x \in a^S$, by Proposition 8.2(b). Thus $|x| = |A| + 2$ and $|x|$ is even. The reflection y^* fixes D and no other point on x, but interchanges the points X, X^y for $X \neq D$ on x. Hence $|x|$ is odd, a contradiction. Our assumption that there exists a central line is therefore false, and (9) is proved.

(10)   *Given any line* x, *there exists a point in* A^G *not on* x.

Let x be any line. If $x \notin A$, we are done. Thus we may assume $x \in A$. If there exists no line y such that $y \notin A$ and $xy \notin J$, then, in particular, $B \subset x^\perp$ and $C \subset x^\perp$, and x would be a central line by Theorem 5.4. This contradicts (9). Thus there is a line y such that $y \notin A$ and $xy \notin J$, and we obtain $x \notin A^y$, as required.

As an immediate consequence of (10) we obtain

(11)   *For all lines* x *we have* $|A| \leq |x| \leq |A| + 2$.

Next we show

(12)   *There exists no completely connected point.*

Suppose that E is some completely connected point. If E is a proper point, then Theorem 8.5 implies that $E(G, S)$ is a Euclidean plane or a hyperbolic-metric plane. The first is impossible, since Axiom $\Delta 1$ holds, and the second is impossible, since the proof of Theorem 8.5 (Case B) shows that every 1-Δ-connected point is contained in a set $x^\perp$, in contradiction to the assumption of our Case B. The case that E is not a proper point is not possible: We would have $xy = yx$ for all x, y in E and $|E| \geq |A| + 2 \geq 4$ by (11) and Lemma 8.4. There would exist a line b not in E and a line c in E, E^b. Then $c \perp b$ and $E \backslash \{c\} \subset c^\perp$, and c would be a central line by Theorem 5.4, in contradiction to (9). This concludes the proof of (12).

(13)   *If* $D \in U(A) \cap A^G$, *then any line through* A *is contained in exactly one point* $E \neq A$ *which is not connected with* D. *For that point* E *we have* $|E| = |A|$.

Let b be any line in A. If $D \cap b^{\perp} \neq \varnothing$ and $c \in D \cap b^{\perp}$, then $c \notin A$ and therefore $A^c \neq A$. Since $b \in A$, A^c and A, $A^c \in U(D)$, we may put $E := A^c$ and obtain $|E| = |A|$.

If $D \cap b^{\perp} = \varnothing$, then Lemma 5.10 yields that there exists a point E on b such that $E \in U(D)$ and $E^{D^2} = E$. If E is Δ-connected, then E is 2-Δ-connected, and Lemma 6.6 implies $D \subset x^{\perp}$ for some suitable x, in contradiction to the assumption made in Case B. Thus E is not Δ-connected. This means, in particular, that $|E| \leq |A|$ by Lemma 8.4 and (11). On the other hand, the map

$$\mu : \begin{cases} D^2 \to E \\ \alpha \mapsto b^{\alpha} \end{cases}$$

is injective (compare the proof of Lemma 5.10). Thus we obtain $|A| = |D| = |D^2| \leq |E| \leq |A|$ and therefore $|E| = |A|$, as required. Since D is Δ-connected, the point C is uniquely determined.

We remark that the case $D \cap b^{\perp} = \varnothing$ in the proof of (13) cannot occur, as we prove in the following.

From now on we separate Case B into part (α) and part (β).

First, we look at *Case B(α)*. Then $A^G \cap U(A) \neq \varnothing$.

(14)   *Any line through A is contained in at least one 2-Δ-connected point.*

Let b be any line and $D \in A^G \cap U(A)$. By (13) any line x through A is contained in exactly one point $X \neq A$ not connected with D. Let $C(A)$ be the set of all these points X. Since D is Δ-connected, any two distinct points in $C(A)$ are not connected by a line. Now observe that $|D| = |A|$, $|C(A)| = |A|$, and $|X| = |A|$ for all X in $C(A)$. Hence by (13) we obtain

$$|S| \geq |\{x \mid x \in X \in (C(A) \cup \{D\})\}| = |A|(|A| + 1) \tag{3}$$

Let E be the uniquely determined point in $C(A)$ on b and put

$$s := \max_{b \in X} |X|$$

Counting S by the intersection of lines with b and observing that $|b| = |A| + 2$ (since $b \notin D$ and $|D| = |A|$ and $D \cap A = \varnothing = D \cap E$), we see that

$$|S| \leq |A| + |E| + |A|(s - 1) - 1 = |A| \cdot s + |A| - 1 \tag{4}$$

Equations 3 and 4 together yield $s > |A|$. From (11), (12), and Lemma 8.4 we deduce $s \leq |A| + 1$ and therefore $s = |A| + 1$. Thus there exists at least one point F on b such that $|F| = |A| + 1$. By Lemma 8.4(e) and (11) F is 2-Δ-connected.

In a further case distinction we now treat Case Bα 1. Then $A^G \cap U(A) \neq \varnothing$ and $|A|$ is even.

(15) $b^{\perp}$ *is a 2-Δ-connected point for all b in A.*

Let b be any line through A and $D \in A^G \cap U(A)$. Then from (13) we deduce the existence of a point $E \neq A$ on b such that $|E| = |A|$. From Proposition 8.2(a) we see that there are lines c, e in A and E, respectively, such that $c, e \in b^\perp$. Obviously, $S(ce) = b^\perp$ and D is connected with $S(ce)$ by a line d, since b, c, e is a triangle and D is Δ-connected. Hence $E = A^d \in A^G$.

Since b is an arbitrary line in A, we have proved that $D \cap x^\perp \neq \emptyset$ for all x in A and therefore $C(A) \subset A^D$. Moreover, if $K \neq E$ is a point in $C(A)$, then we may apply to K and E the result obtained for D and E. Hence $K \cap b^\perp \neq \emptyset$ for all $K \neq E$ in $C(A)$. For $K \neq L$ and $K, L \in C(A)$ the lines in $K \cap b^\perp$ and $L \cap b^\perp$, respectively, are distinct and not equal to c. Thus we obtain $|b^\perp| \geq |A|$. Clearly, $|b^\perp|$ is odd, as otherwise there would exist a line g in $b^\perp \cap c^\perp$ by Corollary 8.3, and we would obtain $bcg = 1$ and $A = g^\perp$, a contradiction to the general assumption made in Case B. Thus $|b^\perp| = |A| + 1$, and $b^\perp$ is a 2-Δ-connected point by (11) and Lemma 8.4.

(16)   *Any line* b *in* A *is contained in* $|A| + 1$ *points of* A^G *and in exactly one* 2-Δ-*connected point* F, *and we have* $F = x^\perp$ *for all* x *in* $b^\perp$.

Obviously, we have $b \in A^x$ for all $x \in b^\perp$. By the proof of (15) we obtain $A^{b^\perp} = \{S(bx) \mid x \in b^\perp\}$, and therefore $|A^{b^\perp}| = |b^\perp| = |A| + 1$. Hence $|A| + 1$ points on b belong to A^G. However, by (14) there exists a 2-Δ-connected point F on b, which is—clearly—not in A^G. From $|b| \leq |A| + 2$ our first assertion follows.

To prove the second part of (16), let x be a line in $b^\perp$. Then $b \in F^x$ and $F^x = F$, since F is the only 2-Δ-connected point on b. Moreover, we see that $x \notin F$, as otherwise $|F|$ would be even and therefore $|A|$ would be odd. Hence $F \subset x^\perp$, and thus $F = x^\perp$ by Theorem 5.4 and (9).

(17)   *In Case* Bα 1 *the S-group plane* E(G, S) *is a Minkowskian plane.*

First, we show that $E(G, S)$ is a complete S-group plane. Let X be any point and let x be a line in X. We choose two lines b, c in A such that $b \perp c$. Then $b^\perp \cap c^\perp = \emptyset$, and therefore $x \notin b^\perp$ or $x \notin c^\perp$. Hence $S(xb) \in A^G$ or $S(xc) \in A^G$ by (16). Another application of (16) for the point $S(xb)$ or $S(xc)$ instead of A shows that x is contained only in Δ-connected points. Hence X is Δ-connected. As we know, $E(G, S)$ contains a quadrilateral.

Now we show that $E(G, S)$ is a Minkowskian plane: Clearly, Axiom $\Delta 1$ is valid. Moreover, Axiom EM holds, since $b^\perp$ and $c^\perp$ are two 2-Δ-connected points not connected by a line [compare (15)].

Next we consider *Case* Bα 2. Then $U(A) \cap A^G \neq \emptyset$ and $|A|$ is odd.

Given any 2-Δ-connected point F, then $|F| = |A| + 1$; this follows immediately from Lemma 8.4(c) and $|a| = |A| + 2$, since $a \notin F$ or $a^s \notin F$ for a suitable s in A.

(18)   *Let F be a 2-Δ-connected point on the line* b *in* A. *Then* $|X|$ *is odd for any point* $X \neq F$ *on* b.

Suppose there would exist a point $H \neq F$ on b such that $|H|$ is even. Then we deduce from Proposition 8.2(a) that there are lines f, h in F and H, respectively, such that $f, h \in b^\perp$. Then $b^\perp = S(fh)$ by Theorem 5.4 and (9). Moreover, $|h|$ is. even, since the reflection b^* fixes the points $b^\perp$ and H on h and interchanges the distinct points X, X^b on h with $X \neq b^\perp$, H. Now $|A| \leq |h| \leq |A| + 2$, and $|A|$ is odd, implying that $|h| = |A| + 1$. Thus F is connected with all the points on h (as $h \notin F$ and $|F| = |A| + 1$). This contradicts Corollary 5.9. Thus our assumption is false, and (18) is proved.

(19)   *Any two distinct 2-Δ-connected points* F, H *are not connected by a line.*

Suppose there exists a line c in $F \cap H$. Proposition 8.2(a) then shows that we can find lines f, h in F and H, respectively, such that $f, h \in c^\perp$. By Corollary 5.9 there exists a point K in $U(F)$ on h. Then $K^c = K$ and $K = c^\perp$ by Theorem 5.4 and (9). Hence $f \in K$ and $F \cap K \neq \varnothing$, a contradiction. This proves (19).

(20) $b^S = S$ *for all* b *in* A.

Let x be any line in S. If $x = b$, then $x \in b^S$. Thus we may assume $x \neq b$. We denote the uniquely determined 2-Δ-connected point on b by F [compare (14) and (19)]. If $S(bx) \neq F$, then $|S(bx)|$ is odd by (18), and our assertion follows from Proposition 8.2(b). If $S(bx) = F$, then we choose a line $y \neq b$ in A. Clearly, $x \neq y$ and $S(xy)$ is not 2-Δ-connected by (19). Thus (18) yields that $|S(xy)|$ is odd. Hence $y \in b^A$ and $x \in y^G$, applying Proposition 8.2(b), and we obtain $x \in b^G$, as required.

(21)   *For any 2-Δ-connected point* F *we have*

$$F = x^\perp \cup \{x\} \qquad \forall x \in F$$

Let b be any line in F. By (20) we may assume that $b \in A$. We put

$$M := b^\perp \cup \{b\} \quad \text{and} \quad N := F \backslash M$$

From (18) we obtain $M \subset F$. Also $|M| \geq 2$, applying Proposition 8.2(a). Our assertion is $N = \varnothing$.

If $x \in N$, then $A \cap A^x = \varnothing$, since $q \in A \cap A^x$ would imply $x \in A$ or $x \notin A$ and $q \perp x$. The first yields $x = b$, a contradiction to $x \notin M$. From the second we deduce $|S(qx)|$ is even, and therefore $S(qx) = F$ by (18). Thus $q = b$ and $x \in b^\perp \subset M$, a contradiction.

If $x \in N$, then $A \cap x^\perp = \varnothing$ (as $A \cap A^x = \varnothing$) and we can apply Lemma 5.10 and obtain that there exists a point D in $U(A)$ on x such that $D^{A^2} = D$. Since any point in A^G is 1-Δ-connected, we see that D does not belong to A^G.

If $x \in N$, then $Mbx \subset N$, since $Mbx \subset F$ and $b^{Mbx} = b^x \neq b$. Obviously, $b^x = b^{ybx} \in A^{ybx} \in A^{Mbx}$ for all y in M, and therefore A^{Mbx} is a set of points in $U(A)$ on b^x. Thus $|A^{Mbx}| \leq 2$. Since $Mbx \subset F$ and therefore $|A^{Mbx}| = |Mbx| = |M|$, we obtain $|M| = 2$. This implies that all the points in $U(A)$ on b^x lie in A^G. From the remark given just above it follows that $b^x \notin N$. Clearly $b^x \neq b$ and therefore $b^x \in b^\perp$. Since $|b^\perp| = 1$, we obtain $|b^N| \leq 1$ and therefore $|N| \leq 2$, applying our remarks in Section 1.

Thus the assumption $N \neq \varnothing$ implies $|M| = |N| = 2$ and $|F| = 4$ and $|A| = 3$. Let b, c, d, e be the lines in F such that $\sigma := bc = de \in J$ and $d, e \in N$. Moreover, let D be the point in $U(A)$ on d such that $D^{A^2} = D$. Then $D^\sigma = D^e \neq D$ and $A \cap D^\sigma = \varnothing$: From $x \in A \cap D^\sigma$ we deduce $x^\sigma \in D$ and therefore $(x^\sigma)^{bx} \in D$, since $bx \in A^2$. However, $x^{\sigma bx} = x^{cx} = c$, since $|S(cx)| = 3$ [as we see from Lemma 8.4, (11), (12), (18), and (20)]. This implies $c \in D$, a contradiction. Thus D, D^σ are two distinct points $\neq F$ on d not in A^G but in $U(A)$. Since A, A^σ are two distinct points in A^G on b, any line is contained in exactly two points in A^G by (20). Considering a line $y \neq b$ in A, we see that $S(dy), S(ey) \in A^G$, since $D, D^\sigma, D^b, D^{\sigma b} \in U(A) \backslash A^G$. Hence y is contained in three mutually distinct points in A^G, a contradiction.

Thus our assumption $N \neq \varnothing$ is false, and so the assertion of (21) is true.

(22)   *Any line is contained in $|A| + 1$ points in A^G and in exactly one 2-Δ-connected point.*

By (20) it suffices to prove the assertion for a line b in A. By (14) and (18) there exists exactly one 2-Δ-connected point F on b. From (21) we see that $F = b^\perp \cup \{b\}$. Hence $b \in A^x$ for all x in F. From $A^x = A^y$ for $x, y \in F$ we deduce $x = y$. Thus we have $|A^F| = |F| = |A| + 1$, and our assertion is proved.

(23)   *In Case Bα 2 the S-group plane* $E(G, S)$ *is a Minkowskian plane.* $E(G, S)$ is a complete S-group plane, since by (22) any point is Δ-connected, and since there exists a quadrilateral in $E(G, S)$. Moreover, $E(G, S)$ is a Minkowskian plane: Clearly, Axiom $\Delta 1$ is valid. From (22) and (19) we see that there exist two 2-Δ-connected points not connected by a line. Hence Axiom EM is fulfilled, and $E(G, S)$ is a Minkowskian plane by definition.

Finally, we wish to investigate *Case Bβ.*

In this case we have $A^G \cap U(A) = \varnothing$, and therefore any two points in A^G are connected by a line.

For any line x we put

$$S_x := \begin{cases} x^\perp & \text{if} \quad |A| \quad \text{is even} \\ x^\perp \cup \{x\} & \text{if} \quad |A| \quad \text{is odd} \end{cases}$$

We prove

(24) $A \cap S_x \neq \emptyset$ *for all lines* x.

In the case that $x \in A$ our assertion follows from Corollary 8.3 if $|A|$ is even, and by definition if $|A|$ is odd. Let $x \notin A$. Then $A \cap A^x \neq \emptyset$, and for $y \in A \cap A^x$ we obtain $y \in x^\perp$ and therefore $y \in A \cap S_x$. This proves (24).

By (24) there exists a line b in $A \cap S_a$. Obviously, $a \neq b$ and therefore $b \in A \cap a^\perp$. In the following the symbol b shall always refer to this line of $A \cap a^\perp$. We have $|A \cap b^\perp| \leq 1$, as otherwise b would be a central line (as $a \notin A$ and $a \in b^\perp$), a contradiction to (9).

(25) *For any line* x *in* A *the set* S_x *is contained in a uniquely determined point* $\bar{S}_x$. *If* $|A|$ *is even, then* $|\bar{S}_x|$ *is odd, and if* $|A|$ *is odd, then* $|\bar{S}_x|$ *is even.*

First, let $|A|$ be even. Then Corollary 8.3 yields the existence of a line y in $A \cap x^\perp$. From (10) we deduce that there exists a point D in A^G not on y, and from (24) we see that there is a line z in $D \cap x^\perp$. Clearly, we have $y \neq z$. If $z \in A$, then $|A \cap x^\perp| \geq 2$, in contradiction to $|A \cap b^\perp| \leq 1$ (compare the introduction in Section 1). Thus $z \notin A$ and $xyz \notin S$. Obviously, we have $S(yz) \subset x^\perp$, and therefore $S(yz) = x^\perp$ by Theorem 5.4 and (9). Thus we may put $\bar{S}_x := S(yz)$ and obtain $S_x = \bar{S}_x$ if $|A|$ is even. Obviously, $|\bar{S}_x| = |S_x|$ is odd, as otherwise there would be lines u, v in S_x such that $uv \in J$, and we would obtain $uvx = 1$ and $A \subset (yuv)^\perp$, a contradiction.

If $|A|$ is odd, then it suffices to prove our assertion for the line b in A [compare Proposition 8.2(b)]. We put $\bar{S}_b := S(ab)$. Then $|\bar{S}_b|$ is even by Proposition 8.2(a). To prove $S_b \subset \bar{S}_b$, let y be any line in S_b. If $y \notin \bar{S}_b$, then $y \neq a$ and $S(ay) \subset b^\perp$, and therefore $S(ay) = b^\perp$ by (9) and Theorem 5.4. From Corollary 8.3 we see that $A \cap b^\perp = \emptyset$, since $|A|$ is odd. Hence without loss of generality we may assume $S(ay) = b^\perp = B$. This yields $B^b = B$ and therefore $C^b = C$ and $C \subset b^\perp$. From that comes $C = b^\perp = B$, a contradiction. Thus we have $y \in \bar{S}_b$ and therefore $S_b \subset \bar{S}_b$. Obviously, $\bar{S}_b$ is uniquely determined.

(26) $|S_x|$ *is the number of points in* A^G *on* x *for all* x *in* A. *Moreover, for any two lines* x, y *in* A *we have* $|S_x| = |S_y|$.

Let x be any line in A and let M_x denote the set of points in A^G on x. Then $A^{S_x} \subset M_x$ and $|A^{S_x}| = |S_x|$. This follows from our remark at the beginning of Case B, since $S_x \subset \bar{S}_x$, $A \cap \bar{S}_x \neq \emptyset$, and $A \neq \bar{S}_x$ by (25). This yields $|S_x| \leq |M_x|$. Let x, y denote two distinct lines in A. If $xy \neq yx$, then $x \notin \bar{S}_y$ and $y \notin \bar{S}_x$. Applying (24), we see that the points in M_x are all connected with S_y by mutually distinct lines and the points in M_y are all connected with S_x by mutually distinct lines. Hence we obtain

$$|S_x| \leq |M_x| \leq |S_y| \leq |M_y| \leq |S_x|$$

This yields $|S_x| = |M_x| = |S_y| = |M_y|$, as required.

If $xy = yx$ and $x \neq y$, then $|A|$ is even. There exists a line z in A such that $xz \neq zx$ and $yz \neq zy$: Let $s \neq x$ be a line in $y^{\perp}$ and $t \neq y$ a line in $x^{\perp}$. Then we have $y \in A^s$ and $x \in A^t$ and $A \neq A^s, A^t$. By the general assumption of Case Bβ, there exists a line $p \in A^s \cap A^t$. We see that $p \notin x^{\perp}, y^{\perp}$, and therefore $p^s \notin x^{\perp}$, $y^{\perp}$ and $p^s \in A$. Thus we may put $z := p^s$.

From $xz \neq zx$ and $yz \neq zy$ we deduce from the arguments given above that $|S_x| = |M_x| = |S_z| = |M_z| = |S_y| = |M_y|$, as required.

(27) $|A^G| = |S_b|^2$.

Let c be any line in A and not in S_b. Then by (26) c is contained in $s := |S_b|$ points of A^G on b. Moreover, from (26) we deduce that any line in a point D of A^G on c is contained in exactly s points of A^G. In particular, this is true for the line joining D and S_b, whose existence follows from (24). Thus each of the s joining lines of points in A^G on c to S_b is contained in s points of A^G. These joining lines are mutually distinct. Hence $|A^G| \geq s^2$. On the other hand, any point in A^G contains a line in S_b by (24). Thus we see that $|A^G| = s^2$.

(28)   *The Case Bβ is not possible.*

Suppose that $U(A) \cap A^G = \varnothing$. We choose a point D in A^G not on b. Such a point exists by (10). From Corollary 5.9 we see that there exists a point H in $U(D)$ on b. Since $U(A) \cap A^G = \varnothing$, we obtain $H \notin A^G$. Let c be a line $\neq b$ in H; then $bc \neq cb$: If $bc = cb$, then $c \in b^{\perp}$. If $|A|$ is even and $d \in A \cap b^{\perp}$, then $c = d^s$ and $s \in b^{\perp}$ by Proposition 8.2(b) and (25). This implies $H = S(bc) = S(bd^s) = S(bd)^s = A^s \in A^G$, a contradiction. If $|A|$ is odd, then $S(bc) = \bar{S}_b$ and $D \cap S_b = \varnothing$, since $D \cap H = \varnothing$. This contradicts (24).

From $bc \neq cb$ we see that $b \notin A^c$ and $b^c \in A^c$, H. Thus A^c is connected not only with all the $|S_b|$ points of A^G on b (since $U(A) \cap A^G = \varnothing$), but also with the two distinct points H and $\bar{S}_b$ on b if $|A|$ is odd, or with H and $\bar{S}_d$ where $d \in A \cap S_b$ if $|A|$ is even. Hence $|A^c| \geq |S_b| + 2$.

On the other hand, recalling (26) and remembering that $U(A) \cap A^G = \varnothing$, we can count A^G from A to obtain

$$|A^G| = |A| (|S_b| - 1) + 1$$

From this we conclude, applying (27), that $|S_b| = |A| - 1$. Thus we obtain $|A| = |A^c| \geq |S_b| + 2 = |A| + 1$, a contradiction.

Thus our assumption is false, and (28) is proved.

From the statements (8), (17), (23), and (28) we see that the assertion of Theorem 8.8 is true. ∎

From the proof of Theorem 8.8 and, in particular, from (8), (17), (23), and Propositions 6.41 and 6.42, we immediately deduce:

COROLLARY 8.9. *Let* E(G, S) *be a finite S-group plane with at least one* 1-Δ-*connected point, say* A. *Then*

(a) E(G, S) *is a hyperbolic-metric plane if and only if there exists a line* a *such that* $A \subset a^{\perp}$

(b) E(G, S) *is a Minkowskian plane if and only if there is no line* a *such that* $A \subset a^{\perp}$.

(c) E(G, S) *is a complete S-group plane of char* $\neq 2$ *if* $|A|$ *is even.*

(d) E(G, S) *is a complete S-group plane of char* 2 *if and only if* $|A|$ *is odd.*

4. REMARKS

It must be remarked that there exist many more types of finite S-groups than those classes presented and studied in this chapter. To give the reader an impression of what can occur in the theory of finite S-groups, we mention here without proof three results due to A. Häussler and U. Ott (compare references 23, 43, and 44). The problem of determining all finite S-groups has not yet been solved. However, Ott classified the structure of any finite S-group (compare reference 44).

In addition to Theorems 8.5 and 8.8 we refer to a result of Häussler in reference 23:

THEOREM 8.10 (Häussler). *Given any finite S-group plane* E(G, S) $\neq$ E(G$_1$, S$_1$), *the following are equivalent*:

(a) E(G, S) *contains at least two distinct* 2-Δ-*connected points connected by a line* (*Axiom H*).

(b) E(G, S) *is a Strubecker plane or a hyperbolic-metric plane or is isomorphic to the group plane over* (PSL(2, K), J) *for a finite field* K *with* $|K| \equiv 1 \mod 4$ *or is isomorphic to the group plane over* (PGL(2, K), S) *for a finite field* K *with* $|K| \equiv 3 \mod 4$ *and* $|K| > 3$, *and* S := {x | Det x *is the quadratic residue class in* K *containing* -1}.

For the definition of the groups $PGL(n, K)$ and $PSL(n, K)$ and for the definition of Det x if x is an element in $PGL(2, K)$ compare the Appendix. We also refer to the Appendix for a description of the groups occurring in the following theorems.

A result concerning finite S-group planes with at least one 2-Δ-connected point analogous to Theorems 8.5 and 8.8 has not yet been found.

THEOREM 8.11 (Ott). *Given any finite S-group plane* E(G, S), *the following are equivalent*:

(a) E(G, S) *contains a polar triangle and a quadrilateral.*

(b)   *The S-group* (G, S) *is isomorphic to one of the following:*
  1. $(PGL(2, K), J)$ *and* K *is a finite field of char* $\neq 2$.
  2. $(PSL(2, K), J)$ *and* K *is a finite field of char* $\neq 2$.
  3. $(PSL(2, K)^*, J)$ *and* $|K| = 9$.
  4. $(PSU(3, K), J)$ *and* $|K| = 16$.

THEOREM 8.12 (Ott).   *Given any finite S-group plane* $E(G, S) \neq E(G_1, S_1)$, *the following are equivalent:*
(a) $E(G, S)$ *contains a central line and two lines* a, b *such that* a, b, ab $\notin Z(G)$.
(b)   *The factor group* $G/Z(G)$ *is isomorphic to one of the following:*
  1. $PGL(2, K)$ *and* K *is a finite field of char* 2 *and* $|K| \geq 4$.
  2. $PSU(3, K)$ *and* K *is a finite field and* $|K| = 2^{2n}$ *and* $n \geq 2$.
  3. $Sz(q)$ *and* $q = 2^{2n+1}$ *and* $n \geq 1$.

We remark that the cases (b)1 in Theorems 8.11 and 8.12 are the cases of the finite hyperbolic-metric planes of char $\neq 2$ (in Theorem 8.11) or of char 2 (in Theorem 8.12).

APPENDIX: AFFINE AND PROJECTIVE PLANES

For the convenience of the reader we first collect some definitions and results on affine and projective planes used in our book but not defined or proved in the text. Clearly, we omit any proof of the results referred to in the following. For that the reader may refer to references 11 or 13.

We then introduce the concept of projective reflection groups for projective planes and elaborate some examples of such projective reflection groups. Moreover, we prove two interesting lemmas on such projective reflection groups.

1. AFFINE PLANES

Let (P, L, I) be a triplet consisting of disjoint sets P and L and a relation $I \subset P \times L$. Then any element in P is called a *point* and any element in L a *line*, and a point A in P is said to be *incident* with the line b in L, if $(A, b) \in I$. If $(A, b) \in I$, we write $A I b$. Moreover, two lines a, b in L are said to be *parallel* if $a = b$ or $a \neq b$ and there does not exist any point A such that $A I a$ and $A I b$. If a is parallel to b, then we write $a \| b$. The triplet (P, L, I) is called an *affine plane* if the following axioms are fulfilled:

(A1) *Given any two distinct points* A, B *in* P. *Then there exists a unique line* c *in* L *such that* $A I c$ *and* $B I c$. *This line* c *will be denoted by* (A, B).

(A2) *Given any point* A *in* P *and any line* b *in* L. *Then there exists a unique line* c *in* L *such that* $A I c$ *and* $b \| c$.

(A3) *There exist at least three points* A, B, C *in* P *such that there does not exist a line* d *with* A, B, C $I d$.

Given an affine plane (P, L, I), we can easily prove the following: If L contains a line a such that a is incident with exactly n points in P, then every line in L is incident with exactly n points. Moreover, every point is incident with exactly $n + 1$ lines, and there exist n^2 points and $n(n + 1)$ lines. The number n is called the *order of the affine plane*. Obviously, $n \geq 2$. If $n = 2, 3, 4, 5, 7, 8, 9$, there exists an affine plane of order n. There is no affine plane of order 6. It is unknown whether an affine plane of order 10 exists.

Statements (D) and (P) are known under the names "affine theorem of Desargues" and "affine theorem of Pappus" (or "affine theorem of Pappus-Pascal"). We emphasize that (D) or (P) need not be true in an arbitrary affine plane.

STATEMENT (D).   *Let* l_0, l_1, l_2 *denote three mutually distinct lines with a common point* O. *The points* P_k, Q_k *are incident with* l_k *and distinct from* O; *k* = 0, 1, 2. *Assume*

$$(P_0, P_1) \| (Q_0, Q_1) \quad and \quad (P_0, P_2) \| (Q_0, Q_2)$$

Then $(P_1, P_2) \| (Q_1, Q_2)$ *(see Figure 1).*

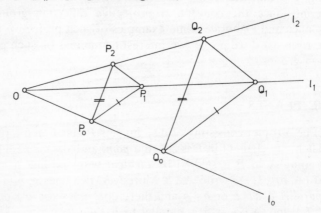

Figure 1

STATEMENT (P).   *Let* l_1 *and* l_2 *denote two distinct lines. Suppose the points* A_k, B_k, C_k *are incident with* l_k *but not with the other line,* $k = 1, 2$. *Let* $(A_1, B_2) \| (A_2, B_1)$ *and* $(A_1, C_2) \| (A_2, C_1)$. *Then* $(B_1, C_2) \| (B_2, C_1)$ *(see Figure 2).*

An affine plane is called a *desarguesian affine plane* if statement (D) holds; it is called a *pappian affine plane* if (P) is true. Hessenberg showed

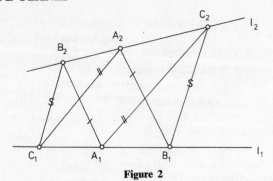

Figure 2

in 1905 that (P) implies (D). Thus any pappian affine plane is a desarguesian plane.

Given any skew-field K, we can construct an affine plane over K as
follows: Let $V := K^2 = \{(x, y) \mid x, y \in K\}$. Then V is a vector space over K.
Let L be the set of residue classes of V modulo one-dimensional
subspaces of V, and let $\in$ denote the inclusion relation of vectors in
elements of L. Then $(V, L, \in)$ is a desarguesian affine plane. It is called
the *affine coordinate plane* over V (or over K) and is denoted by $A(V, K)$.
For brevity, we may also write $A(V)$ or $A(K)$ instead of $A(V, K)$.

If K is a field, then $A(K)$ is a pappian affine plane.

It is obvious how to define isomorphisms of affine planes. The so-called
coordinization theorem implies that any desarguesian affine plane is
isomorphic to the affine coordinate plane over a suitable skew-field and
that any pappian affine plane is isomorphic to the affine coordinate plane
over a field.

2. PROJECTIVE PLANES

In any affine plane we have to distinguish between pairs of parallel and
pairs of intersecting lines. However, any affine plane can be embedded
into a system in which this distinction vanishes. We do this as follows:

We start with two disjoint sets P of points and L of lines and an
incidence relation $I \subset P \times L$. Capital and lowercase letters denote points
and lines, respectively. The triplet (P, L, I) is called a *projective plane*, if it
satisfies the following conditions:

(P1)   *Given any pair of points there exists a line incident with both.
Given any pair of lines there exists a point incident with both.*

(P2)   *If both of the points* A, B *are incident with both of the lines* c, d, *then*
$A = B$ *or* $c = d$.

(P3)   *There exist four points no three of them incident with the same line.*

If A, B are two distinct points in a projective plane, then (A, B) denotes the uniquely determined line c such that A, $B \, \mathrm{I} \, c$. If a, b are two distinct lines in a projective plane, then (a, b) or $\cdot \, a \cdot b$ denotes the uniquely determined point C such that $C \, \mathrm{I} \, a$, b. A set of three points not incident with the same line is called a *triangle* and a set of three nonconcurrent lines is called a *trilateral*. A set of four lines is called a *quadrilateral* if no three of the lines are concurrent. If a, b, c, d is a quadrilateral, then the points $a \cdot b$, $b \cdot c$, $c \cdot d$, $d \cdot a$ are called the *vertices* of the quadrilateral, and the points $a \cdot c$, $b \cdot d$, $(a \cdot b, c \cdot d) \cdot (b \cdot c, d \cdot a)$ are called the *diagonal points* of that quadrilateral.

Given any projective plane (P, L, I) and a line a in L which is incident with exactly $n + 1$ points, then any line is incident with exactly $n + 1$ points and any point is incident with exactly $n + 1$ lines. Moreover, the projective plane contains exactly $n^2 + n + 1$ points and $n^2 + n + 1$ lines. The number n is called the *order of the (finite) projective plane*.

The projective theorem of Desargues reads:

STATEMENT (D)*. *If two triangles are perspective from a center, they are also perspective from an axis (see Figure 3).*

Perspectivity of triangles can be defined as follows: Let $A_0 A_1 A_2$ and $B_0 B_1 B_2$ denote two triangles and put

$$p_0 := (A_1, A_2), \qquad p_1 := (A_2, A_0), \qquad p_2 := (A_0, A_1)$$

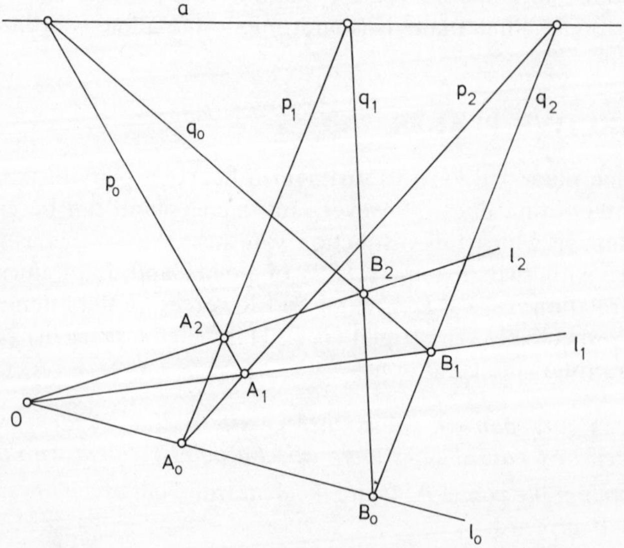

Figure 3

and

$$q_0 := (B_1, B_2), \qquad q_1 := (B_2, B_0), \qquad q_2 := (B_0, B_1)$$

Then $A_0 A_1 A_2$ and $B_0 B_1 B_2$ are perspective from the center O if the points O, A_k, B_k are collinear; they are perspective from the axis a if the lines a, p_k, q_k are concurrent; $k = 0, 1, 2$.

The projective theorem of Pappus can be formulated as follows:

STATEMENT (P)*. *Suppose no two consecutive points of the cyclic sequence*

$$\ldots A_1, B_2, C_1, A_2, B_1, C_2, A_1, \ldots$$

are collinear with any two of the remaining ones, but the three points A_k, B_k, C_k *are collinear;* $k = 1, 2$. *Then the three points*

$$A_0 = ((B_1, C_2), (C_1, B_2)), \qquad B_0 := ((C_1, A_2), (C_2, A_1)),$$
$$C_0 := ((A_1, B_2), (A_2, B_1))$$

are collinear (see Figure 4).

A projective plane is called a *desarguesian* or a *pappian projective plane* if Statement (D)* or (P)* is valid, respectively. As in the affine case, any pappian projective plane is a desarguesian plane.

Given any skew field K, a projective plane over K can be constructed as follows: Let $V := K^3 = \{(x, y, z) \mid x, y, z \in K\}$, and let P and L denote the set of all one-dimensional subspaces of V or the set of all two-dimensional subspaces of V, respectively. Finally, we denote the inclusion

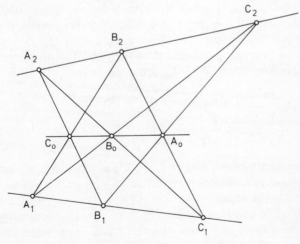

Figure 4

relation by $\subset$. Then $(P, L, \subset)$ is a desarguesian projective plane. We call it the *projective coordinate plane* over V (over K), and denote it by $\Pi(V, K)$ or, briefly, by $\Pi(V)$ or by $\Pi(K)$. An isomorphic model of the projective coordinate plane over a skew-field is obtained if we interchange the sets P and $L : (L, P, \supset)$ is as good a projective plane as $(P, L, \subset)$.

$\Pi(V, K)$ is a pappian projective plane if and only if K is a field.

The coordinate field K is of characteristic $\neq 2$ if for any quadrilateral the diagonal points are noncollinear. If $\Pi(V, K)$ contains at least one quadrilateral whose diagonal points are noncollinear, then K is of characteristic $\neq 2$. Conversely, K is of characteristic 2 if any quadrilateral has collinear diagonal points.

A desarguesian projective plane Π is called a *desarguesian projective plane of char$\neq 2$* if any quadrilateral has noncollinear diagonal points. Otherwise Π is called a *desarguesian projective plane of char 2*.

Another representation of the projective coordinate plane over a field K can be obtained if we put

$$\mathrm{I} := \{((x_1, x_2, x_3), (u_1, u_2, u_3)) \mid x_1 u_1 + x_2 u_2 + x_3 u_3 = 0\} \subset P \times P.$$

Then (P, P, I) is a projective plane isomorphic to $\Pi(V, K)$.

3. PROJECTIVE CLOSURE OF AN AFFINE PLANE

Let (P, L, I) be an affine plane. Since the relation in L

$$a \quad \text{is parallel to} \quad b$$

is an equivalence relation, we have a decomposition of L into classes of parallel lines. These classes are called *parallel pencils*. If a is a line in L, we denote the parallel pencil containing a by Π_a. Any parallel pencil is called an *improper point*, and the set P_ω of all improper points is called an *improper line*, denoted by l_∞. Then we define

$$P^* := P \cup P_\omega$$

$$L^* := L \cup \{l_\infty\}$$

$$\mathrm{I}^* := \mathrm{I} \cup \{(\Pi, l_\infty) \mid \Pi \in P_\omega\} \cup \{(\Pi, l) \mid \Pi \in P_\omega \quad \text{and} \quad l \in \Pi\}$$

We can easily prove that (P^*, L^*, I^*) is a projective plane. That projective plane is called the *projective closure of the affine plane* (P, L, I). Obviously the affine plane (P, L, I) is embedded into the projective plane (P^*, L^*, I^*), and for (P^*, L^*, I^*) there exists no distinction between pairs of parallel lines and pairs of intersecting lines, since any two lines in a projective plane do intersect.

Conversely to the process of embedding an affine plane into a projective plane, we construct for any given projective plane an affine specialization: Let (P, L, I) denote an arbitrary projective plane and let a be any line in L. We put

$$P_a := P \setminus \{X \mid X \, I \, a\}$$

$$L_a := L \setminus \{a\}$$

$$I_a := I \big|_{P_a \times L_a} = I \cap (P_a \times L_a)$$

Then (P_a, L_a, I_a) is an affine plane called the *affine specialization* of (P, L, I) with respect to the line a.

The affine plane (P, L, I) is a desarguesian or a pappian affine plane if and only if its projective closure (P^*, L^*, I^*) is a desarguesian or a pappian projective plane, respectively.

4. AFFINE SPACES

Let K denote any skew-field and let V be any right vector space over K. Any triplet (A, V, Φ) consisting of nonempty sets A and V and a map Φ of $A \times A$ into V is called an *affine space* and the elements in A are called *points* if the following conditions are satisfied:

(AS 1). *Given any three points* X, Y, Z *in* A *then*

$$\Phi(X, Y) + \Phi(Y, Z) = \Phi(X, Z)$$

(AS 2) *Given any point* O *in* A *then the map*

$$\psi : \begin{cases} A \to V \\ X \mapsto \Phi(O, X) \end{cases}$$

   is bijective.

We put

$$\overline{XY} := \Phi(X, Y) \qquad \forall X, Y \in A$$

and call $\overline{XY}$ the *vector* from X to Y.

If $\dim V = n$, then (A, V, Φ) is called an *n-dimensional affine space*.

Let V be any right vector space over the skew-field K. We consider the map

$$\Phi : \begin{cases} V \times V \to V \\ (X, Y) \mapsto X - Y \end{cases} \tag{1}$$

Then (V, V, Φ) is an affine space called the *standard affine space* over V. Up to isomorphisms the standard affine spaces are all the affine spaces.

If (A, V, Φ) is an affine space and O is a point, then for a subspace T of V the subset of A given by

$$U := \{X \mid \overline{OX} \in T\}$$

is called an affine subspace of (A, V, Φ), and is denoted by $U(O, T)$. In particular, if $\dim T = k$, then U is called a *k-dimensional affine subspace* of (A, V, Φ). **Any** zero-dimensional affine subspace contains exactly one point. Identifying this subset of A with the point itself, we can say that the zero-dimensional affine subspaces of (A, V, Φ) are the *points*. The one-dimensional affine subspaces of (A, V, Φ) are called *lines*. Thus we see that the affine coordinate plane $A(K^2, K)$ is isomorphic to the standard affine space (K^2, K^2, Φ) with Φ from equation 1. The two-dimensional affine subspaces of an affine space (A, V, Φ) are called *planes*. Two affine subspaces $U(O, T)$ and $U(O', T')$ of an affine space (A, V, Φ) are said to be *parallel* if $T \subset T'$ or $T' \subset T$.

5. PERSPECTIVE COLLINEATIONS OF A PROJECTIVE PLANE

Let (P, L, I) and (P', L', I') denote two projective planes. A pair $\pi :=$ (α, α') of bijections of P onto P' and of L onto L' is called a *collineation* (or an isomorphism) if the following holds:

$$A \mathrm{I} b \Leftrightarrow A\alpha \mathrm{I}' b\alpha' \qquad \forall A \in P; b \in L$$

A collineation π is called a *perspective collineation* with center O and axis a if

$$x\pi = x \qquad \forall \text{ lines } x \text{ with } O \mathrm{I} x$$

$$X\pi = X \qquad \forall \text{ points } X \text{ with } X \mathrm{I} a$$

Any collineation fixing all points on a line is a perspective collineation. Dual to that, any collineation fixing all lines through a point is a perspective collineation.

Given any pair (A, b) in $P \times L$, then the set of all the perspective collineations with center A and axis b form a group, which we denote by $P(A, b)$. Moreover, the sets

$$P(A) := \bigcup_{b \in L} P(A, b)$$

$$P(b) := \bigcup_{A \in P} P(A, b)$$

$$P(A, B) := \bigcup_{B \mathrm{I} b} P(A, b)$$

$$P(a, b) := \bigcup_{A \mathrm{I} a} P(A, b)$$

form groups.

Any perspective collineation π in $P(A, b)$ is uniquely determined by one pair $(B, B\pi)$ with $A \neq B$ and $B \not\hspace{-0.4em}\mathrm{I}\, b$.

If $A \not\hspace{-0.4em}\mathrm{I}\, b$, then any perspective collineation in $P(A, b)$ is called a *homology*, and if $A \,\mathrm{I}\, b$, then any element in $P(A, b)$ is called an *elation*.

Any product of perspective collineations of a projective plane is called a *projective collineation*. The set of all the projective collineations of a projective plane form a group.

If α is a collineation, then $\alpha^{-1}P(A, b)\alpha = P(A\alpha, b\alpha)$ for all pairs (A, b) in $P \times L$. Hence any two groups $P(A, b)$ and $P(A', b')$ of homologies of a desarguesian projective plane are isomorphic (since there exists a collineation α such that $A' = A\alpha$ and $b' = b\alpha$). Analogously, any two groups $P(A, b)$ and $P(A', b')$ of elations of a desarguesian projective plane are isomorphic.

6. COLLINEATIONS INDUCED BY LINEAR MAPPINGS

Let $\Pi := \Pi(V, K)$ be the projective coordinate plane over a field K. If σ is a linear mapping of V onto itself, then σ maps one-dimensional subspaces onto one-dimensional subspaces and two-dimensional subspaces onto two-dimensional subspaces. Clearly, σ preserves the inclusion relation. Thus σ induces a collineation $\bar{\sigma}$ of Π if we put

$$X\bar{\sigma} := \{x\sigma \mid x \in X\}$$

for all one-dimensional subspaces X of V and

$$U\bar{\sigma} := \{u\sigma \mid u \in U\}$$

for all two-dimensional subspaces U of V.

The group of all the projective collineations of $\Pi(V, K)$ coincides with the group of the collineations of $\Pi(V, K)$ which are induced by linear mappings of V onto itself.

7. PROJECTIVE REFLECTIONS

Let Π be any projective plane. Then any involutory perspective collineation in $P(A, b)$ is called a *projective reflection* in the point A or in the line b, respectively. Sometimes we also call the identity a *projective reflection* (in any point and in any line). We recall some statements on projective reflections. Some of them are only proved in the cited book [15] for the case of char $\neq 2$. The extensions to the case of char 2 can easily be done, and are left as an exercise for the reader.

The product of two projective reflections $\neq 1$ in the same point A but in distinct lines b, c not incident with A is an elation in $P(A, (A, b \cdot c))$. Dually, the product of two projective reflections $\neq 1$ in the same line a but in distinct points B, C not incident with a is an elation in $P(a \cdot (B, C), a)$.

Given any pair (A, b) consisting of a point A and a line b, then there may exist more than one projective reflection $\neq 1$ in the point A and in the line b. However, if $A \not\equiv b$ and if there exists a collineation, fixing A and moving b or fixing b and moving A, then there is at most one projective reflection $\neq 1$ in the point A and in the line b. In particular, in any pappian projective plane of char $\neq 2$ for any given point A and any line b not through A there exists exactly one projective reflection $\neq 1$ in the point A and in the line b.

If σ is a projective reflection in the point A and in the line b and if α is any collineation, then $\alpha^{-1}\sigma\alpha$ is a projective reflection in the point $A\alpha$ and in the line $b\alpha$.

If σ is a projective reflection $\neq 1$ in the line a and τ is an elation $\neq 1$ with axis b, then $\sigma\tau$ is a projective reflection if and only if $a = b$; in this case a is an axis of $\sigma\tau$. Dually, the product of a projective reflection $\neq 1$ in the point A with an elation $\neq 1$ with center B is a projective reflection if and only if $A = B$; for $A = B$ $\sigma\tau$ is a projective reflection in the point A.

The product of three projective reflections $\neq 1$ in the same point A but not in the same line is a projective reflection in the point A. Dually, the product of three projective reflections $\neq 1$ in the same line b but not in the same point is a projective reflection in the line b.

If the product of three projective reflections in the point A and in the lines a, b, c, respectively, is a projective reflection in the line d, then $a = b$ and $c = d$ or $a \neq b$ and (a, b), (c, d), A are collinear.

Now let Π be any desarguesian projective plane. If Π is of char $\neq 2$, then any projective reflection $\neq 1$ is an involutory homology. If Π is of char 2, then a projective reflection is an elation. In the following we do not regard the identity as a projective reflection in a desarguesian projective plane of char $\neq 2$. Hence in any such plane, for every pair consisting of a point A and a line b not through A there exists exactly one projective reflection in the point A and in the line b.

8. PROJECTIVE REFLECTION GROUPS

Let Π denote an arbitrary projective plane. Any group of collineations of Π which is generated by projective reflections is called a *projective reflection group*. Sometimes we also call the pair (G, S) consisting of a set S of projective reflections and the group G generated by S a projective reflection group. We wish to give some examples and to add some remarks if Π is a pappian projective plane.

(a) Trivially, given a projective reflection σ, than it generates a projective reflection group isomorphic to the cyclic group of order 1 or 2.

(b) Let A, B denote two distinct points in Π. We put

$$S(A, B) := \{\sigma \mid \sigma \text{ projective reflection in } P(A, B)\}$$

The group generated by $S(A, B)$ may be denoted by $PR(A, B)$.

If Π is a pappian projective plane, then $PR(A, B)$ contains the group $P(A, (A, B))$ of all the elations with center A and axis (A, B).

Moreover, we have

$$\sigma, \sigma', \sigma'' \in S(A, B) \quad \text{implies} \quad \sigma\sigma'\sigma'' \in S(A, B)$$

(We recall that only in the case of char 2 can it happen that the identity is a projective reflection.)

If Π is a pappian projective plane of char 2, then $S(A, B) = P(A, (A, B)) = PR(A, B)$. If Π is a pappian projective plane of char $\neq 2$, then $PR(A, B)$ is a generalized dihedral group (in the sense of reference 2) with the group $P(A, (A, B))$ as its abelian subgroup of index 2.

(c) Dual to the example (b) we define for any distinct lines a, b in any projective plane Π:

$$S(a, b) := \{\sigma \mid \sigma \text{ projective reflection in } P(a, b)\}$$

The group generated by $S(a, b)$ is denoted by $PR(a, b)$. Obviously, the dual statements to that given above for example (b) are valid for $S(a, b)$ and $PR(a, b)$.

(d) Let A denote a fixed point in a projective plane Π. Then we put

$$S(A) := \{\sigma \mid \sigma \text{ projective reflection in the point } A\}$$

and denote by $PR(A)$ the group generated by $S(A)$.

If Π is a pappian projective plane, then $P(A, A) \subset PR(A)$. Moreover, we have

$$\sigma, \sigma', \sigma'' \in S(A) \quad \text{implies} \quad \sigma\sigma'\sigma'' \in S(A).$$

If Π is of char 2, then $S(A) = P(A, A) = PR(A, A)$. If Π is of char $\neq 2$, then PR(A) is a generalized dihedral group with the group $P(A, A)$ as its abelian subgroup of index 2.

If (V, Q) is a three-dimensional metric vector space over a field K of char $\neq 2$ such that dim rad $(V, Q) = 2$ and Ind $(V, Q) = 0$, then the elements in the set $S(V, Q)$ of reflections of (V, Q) induce projective reflections in the projective coordinate plane $\Pi(V)$. The set of all these induced projective reflections is $S(A_\omega)$ if we denote the point in $\Pi(V)$ given by the two-dimensional subspace $\text{rad}\,(V, Q)$ by A_ω. Hence the group $PR(A_\omega)$ is isomorphic to the group $O(V, Q)$.

(e) Dual to (d) we consider for any given line b in an arbitrary projective plane Π the set

$$S(b) := \{\sigma \mid \sigma \quad \text{projective reflection in the line} \quad b\}$$

and denote by $PR(b)$ the group generated by $S(b)$. Dualizing the statements obtained for example (d) we arrive at the analogous statements for the group $PR(b)$.

If we consider the affine specialization Π_b of Π with respect to the line b as the line at infinity, then the restrictions of the collineations in $PR(b)$ form the group of affine collineations of Π_b generated by the reflections of Π_b in the points of Π_b (reflection in the sense of reference 11, i.e., an involutory collineation of the affine plane fixing all the lines through a fixed point).

(f) Let (A, b) denote a fixed flag in an arbitrary projective plane Π, that is, a pair consisting of a point A and a line b through A.

Then we consider the set

$$S(A, b) := \{\sigma \mid \sigma \quad \text{projective reflection in} \quad P(X, y)$$
$$\text{with} \quad X \,\mathrm{I}\, b \quad \text{and} \quad A \,\mathrm{I}\, y\}$$

$PR(A, b)$ may denote the group generated by $S(A, b)$.

By the remark at the end of Section 5 we know that any two pairs $(PR(A, b), S(A, b))$ and $(PR(A', b'), S(A', b'))$ for any desarguesian projective plane are isomorphic.

If Π is a pappian projective plane, then $P(A, A) \cup P(b, b) \subset PR(A, b)$. Moreover, we show

LEMMA A1. *For any flag* (A, b) *in a pappian projective plane the group* $P(A, b)$ *is the center of the group* $PR(A, b)$.

PROOF. We have the following equivalent statements: $\alpha \in Z(PR(A, b)) \Leftrightarrow \alpha\sigma = \sigma\alpha$ for all σ in $PR(A, b) \Leftrightarrow \alpha^{-1}\sigma\alpha = \sigma$ for all $\sigma \in PR(A, b)$.

Since $P(A, A) \cup P(b, b) \subset PR(A, b)$, we obtain the following for all α in $Z(PR(A, b)): P(X, y) = \alpha^{-1}P(X, y)\alpha = P(X\alpha, y\alpha)$ and therefore $X\alpha = X$ and $y\alpha = y$ for all $X \,\mathrm{I}\, b$ and $A \,\mathrm{I}\, y$. This implies $\alpha \in P(A, b)$ and $Z(PR(A, b)) \subset P(A, b)$.

Conversely, we have $\alpha\sigma = \sigma\alpha$ for all $\alpha \in P(A, b)$ and all $\sigma \in S(A, b)$ (compare reference 15), and therefore $P(A, b) \subset Z(PR(A, b))$. Thus $Z(PR(A, b)) = P(A, b)$, and Lemma A1 is proved. ∎

If Π is a pappian projective plane of char 2, then $S(A, b) = P(A, A) \cup P(b, b)$.

If Π is a pappian projective plane of char $\neq 2$, we prove:

LEMMA A2. *Let* (A, b) *be a flag in the pappian projective plane* Π *of char* $\neq 2$. *Then*[*]

$$PR(A, b)/P(A, b) \cong PR(A).$$

If we denote this isomorphism by φ, *then* $S(A, b)\varphi = S(A)$.

PROOF. We may consider the projective plane Π the projective coordinate plane $\Pi(V, K)$ over a field K of char $\neq 2$. Moreover, we may assume that $V = K^3$ and $b = \langle (1, 0, 0) \rangle$ and that A corresponds to the two-dimensional subspace $\langle (1, 0, 0), (0, 1, 0) \rangle$. Then any projective reflection in $PR(A, b)$ is induced by a linear mapping of V onto itself with the matrix

$$B(u, v) := \begin{pmatrix} -1 & u & \frac{1}{2}uv \\ 0 & 1 & v \\ 0 & 0 & -1 \end{pmatrix} \qquad u, v \in K$$

If u, v range over all the elements in K, then we obtain all projective reflections in $S(A, b)$, and if $(u, v) \neq (u', v')$, then $B(u, v) \neq B(u', v')$ and the projective reflections induced by $B(u, v)$ and $B(u', v')$, respectively, are distinct. The matrix group generated by the matrices $B(u, v)$ $[u, v \in K]$ is the group

$$G := \{ A(s, u, v, w) \mid u, v, w \in K \quad \text{and} \quad s = \pm 1 \}$$

if

$$A(s, u, v, w) := \begin{pmatrix} s & u & w \\ 0 & 1 & v \\ 0 & 0 & s \end{pmatrix}$$

Obviously, we have $G \cong PR(A, b)$. The matrices

$$C(w) := A(1, 0, 0, w) \qquad \text{for} \quad w \in K$$

induce all the elations in $P(A, b)$. We put

$$T := \{ C(w) \mid w \in K \}$$

and obtain $P(A, b) \cong T \cong K^+$ if K^+ denotes the additive group of the field K.

---

[*] The same statement holds if Π is of char 2. However, we do not need this result in this book, and thus we may omit the proof.

On the other hand, any projective reflection in the group $PR(A)$ is induced by a linear mapping of V onto itself with the matrix

$$D(u, v) := \begin{pmatrix} -1 & 0 & u \\ 0 & -1 & -v \\ 0 & 0 & 1 \end{pmatrix} \qquad u, v \in K$$

and $(u, v) \neq (u', v')$ implies that the projective reflections induced by the linear maps with the matrices $D(u, v)$ and $D(u', v')$, respectively, are distinct. The matrix group generated by the matrices $D(u, v)$ $(u, v \in K)$ is the group

$$G' := \{E(s, u, v) \mid u, v \in K \quad \text{and} \quad s = \pm 1\}$$

if

$$E(s, u, v) := \begin{pmatrix} s & 0 & u \\ 0 & s & sv \\ 0 & 0 & 1 \end{pmatrix}$$

Clearly, the groups $PR(A)$ and G' are isomorphic. Now we consider the map

$$\Phi : \begin{cases} G \to G' \\ A(s, u, v, w) \mapsto E(s, u, v). \end{cases}$$

Then Φ is surjective, trivially. Moreover, Φ is a homomorphism:

$$(A(s, u, v, w)A(s', u', v', w'))\Phi = A(ss', su'+u, v'+s'v, sw'+uv'+s'w)\Phi$$
$$= E(ss', su'+u, v'+s'v)$$
$$= E(s, u, v) \cdot E(s', u', v')$$
$$= (A(s, u, v, w)\Phi)(A(s', u', v', w')\Phi)$$

Obviously, $\ker \Phi = \{A(1, 0, 0, w) \mid w \in K\} = T$. Hence we have $G/T \cong G'$ and therefore, since $G \cong PR(A, b)$ and $T \cong P(A, b)$ and $G' \cong PR(A)$,

$$PR(A, b)/P(A, b) \cong PR(A)$$

as required.

The last statement of Lemma A2 is clear. ∎

We recall that in Chapter 6, Section 6 we proved (Proposition 6.17 by Hoffmann) that the projective reflection group $(PR(A, b), S(A, b))$ for a pappian projective plane of char $\neq 2$ is an S-group.

9. DEFINITION OF SOME CLASSICAL GROUPS

We conclude the Appendix by adding the definition of some classical groups mentioned in our book.

Let (V, K) denote any vector space. Then the set $GL(V, K)$ of all the bijective linear maps of V onto itself forms a group, called the *general linear group*. If the dimension of V is finite and dim $V = n$, we sometimes write $GL(n, K)$ or $GL_n(K)$ instead of $GL(V, K)$, since up to isomorphisms there exists exactly one vector space V over K with dimension n. If K is finite and $q := |K|$, then we write $GL(n, q)$ instead of $GL(n, K)$.

The set of all the elements σ in $GL(V, K)$ with determinant 1 forms a subgroup of $GL(V, K)$ and is denoted by $SL(V, K)$. Clearly, we may interchange $SL(V, K)$ by $SL(n, K)$ or $SL_n(K)$ or $SL(n, q)$ if V is finite-dimensional or if K is finite.

For any vector space (V, K) we may define the projective coordinate space as follows: Let R be the set of all the one-dimensional subspaces of V, and L the set of the two-dimensional subspaces of V. Then we call any element in R a *point* and any element in L a *line*, and a point A in R *incident* with a line b in L if $A \subset b$. Then the triplet $(R, L, \subset)$ is called the *projective coordinate space* over the vector space (V, K), denoted by $\Pi(V, K)$. A *collineation* of $\Pi(V, K)$ is defined as a bijective mapping σ of R onto itself which maps any three points on the same line onto three points incident with the same line. If $\sigma \in GL(V, K)$, then σ induces a collineation $\bar{\sigma}$ of $\Pi(V, K)$ by putting

$$\langle X \rangle \bar{\sigma} = \langle X\sigma \rangle \qquad \forall X \neq O \quad \text{in} \quad V$$

All the collineations of $\Pi(V, K)$ induced by a linear map in $GL(V, K)$ form a group, denoted by $PGL(V, K)$. We also may write $PGL(n, K)$ or $PGL_n(K)$ if V is finite dimensional, or $PGL(n, q)$ if K is finite. The subset of $PGL(n, K)$ consisting of all the collineations induced by elements in $SL(V, K)$ forms a subgroup of $PGL(V, K)$ and is denoted by $PSL(V, K)$ [or by $PSL(n, K)$ and $PSL_n(K)$ and $PSL(n, q)$, respectively].

Let (V, K) be any vector space and let μ denote any automorphism $\neq 1$ of the field K such that $\mu^2 = 1$. Then a map $f : V \times V \to K$ is called a *hermitian form* on V if the following properties are fulfilled:

$$f(xA + yB, C) = xf(A, C) + yf(B, C)$$
$$f(A, B) = \mu(f(B, A))$$

for all x, y in K and all A, B, C in V. The hermitian form f is called *nondegenerate* if

$$f(X, A) = 0 \quad \text{for all} \quad X \quad \text{in} \quad V \quad \text{implies} \quad A = O$$

The set of all elements σ in $GL(V, K)$ such that

$$f(X\sigma, Y\sigma) = f(X, Y) \qquad \forall X, Y \in V$$

forms a group, which is called the *unitary group* and which is denoted by $U(V, f)$. The symbols $PU(V, f)$ and $PSU(V, f)$ denote the set of all the collineations induced by elements in $U(V, f)$ or in $U(V, f) \cap SL(V, K)$. It is clear what is meant by $U(n, q)$, $PU(n, K)$, and so forth.

In particular, if $K = GF(q^2)$ is a finite field with $|K| = q^2$ and q is even, then we consider the special automorphism of K given by

$$\tau : \begin{cases} K \to K \\ x \mapsto x^q \end{cases}$$

Then τ is involutory, and up to isomorphism there exists exactly one unitary vector space (V, f) with dim $V = 3$ and a nondegenerate hermitian form f with respect to the automorphism τ. In Chapter 8, Section 4, Theorems 8.11 and 8.12 refer to these special unitary vector spaces and unitary groups.

Finally, we consider the Galois field $K := GF(q)$ such that $q = 2^{2n+1}$ and n is a positive integer ≥ 1. Then there exists exactly one automorphism τ of K such that $x\tau^2 = x^2$ for all x in K. We put $V := K^4$ and

$$O := \{\langle(x, y, z, 1)\rangle \mid x, y \in K \quad \text{and} \quad z = xy + (x\tau)x^2 + y\tau\}$$

Then O is a set of one-dimensional subspaces of V and therefore a set of points in the projective coordinate space $\Pi(V, K)$. Moreover, it has the following properties:

(1) Any line in $\Pi(V, K)$ intersects O in at most two points.

(2) If $A \in O$, then the set of all the points on lines l such that $l \cap O = \{A\}$ is a plane in $\Pi(V, K)$, that is, the set of one-dimensional subspaces of V contained in a fixed three-dimensional subspace of V.

Any set of points in the projective coordinate space $\Pi(V, K)$ fulfilling (1) and (2) is called an *ovoid*. The set of all collineations π in $PGL(V, K)$ such that $\pi(O) = O$ form a group, called a *Suzuki group* and denoted by $Sz(q)$.

BIBLIOGRAPHY

The following books and reports are of special interest in the context of our book.

[1] Artin, E. *Geometric algebra*: Interscience, New York, 1957.

[2] Bachmann, F. *Aufbau der Geometrie aus dem Spiegelungsbegriff*, 2nd ed. Springer-Verlag, Berlin–New York, 1973.

[3] Coxeter, H. S. M. *Non-euclidean geometry*. University of Toronto Press, Toronto, 1957.

[4] Coxeter, H. S. M. *The real projective plane*, 2nd ed. Cambridge University Press, Cambridge, 1960.

[5] Dembowski, P. *Finite geometries*. Springer-Verlag, New York, 1968.

[6] Ewald, G. *Geometry*. Wadsworth, Belmont, 1971.

[7] Hilbert, D. *Grundlagen der Geometrie*, 11th ed., Teubner Verlag, Stuttgart, 1972.

[8] Hughes, D. R. and Piper, F. C. *Projective planes*. Springer-Verlag, New York, 1973.

[9] Karzel, H., Sörensen, K., and Windelberg, W. *Einführung in die Geometrie*. Vandenhoeck & Ruprecht, Göttingen, 1973.

[10] Lam, T. Y. *The algebraic theory of quadratic forms*. Benjamin, New York, 1973.

[11] Lingenberg, R. *Grundlagen der Geometrie*, 3rd ed. Bibliographisches Institut, Mannheim, 1978.

[12] O'Meara, O. T. *Introduction to quadratic forms*. Springer-Verlag, New York, 1963.

[13] Pickert, G. *Projektive Ebenen*, 2nd ed. Springer-Verlag, Berlin–New York, 1975.

[14] Salzmann, H. *Topological planes*. Advances in Mathematics, Vol. 2, No. 1 (1967), 1–60.

[15] Scherk, P. and Lingenberg, R. *Rudiments of plane affine geometry*. University of Toronto Press, Toronto, 1975.

[16] Snapper, E. and Troyer, R. J. *Metric affine geometry*. Academic Press, New York, 1971.

[17] Bachmann, F. Geometrien mit nicht-euklidischer Metrik, in denen es zu jeder Geraden durch einen nicht auf ihr liegenden Punkt mehrere Nichtschneidende gibt. I, II, III, *Mathematische Zeitschrift*, Vol. 51, pp. 752–768, 769–779 (1949). *Mathematische Nachrichten*, Vol. 1, pp. 258–276 (1948).

[18] Bollow, B. Modelle der metrisch-euklidischen Geometrie. *Archiv der Mathematik,* Vol. 20, pp. 94–106, 202–213 (1969).

[19] Ellers, E. and Sperner, E. Einbettung eines desarguesschen Ebenenkeims in eine projektive Ebene. *Abhandlungen aus dem mathematischen Seminar der Universität Hamburg,* Vol. 25, pp. 206–230 (1962).

[20] Gunther, G. Singular isometries in orthogonal groups. To appear in *Canadian Mathematical Bulletin.*

[21] Gunther, G. A geometric characterization of orthogonal groups of characteristic 2. *Geometriae dedicata,* Vol. 5, pp. 321–346 (1976).

[22] Gunther, G. and Nolte, W. Defining relations in orthogonal groups of char 2. To appear in *Canadian Journal of Mathematics.*

[23] Häussler, A. Charakterisierung einer Klasse verallgemeinerter hyperbolisch-metrischer Bewegungsgruppen. Thesis, Karlsruhe, 1975.

[24] Hoffmann, D. Richtungsaffine Ebenen. Thesis, Karlsruhe, 1974.

[25] Karzel, H. Ein Axiomensystem der absoluten Geometrie. *Archiv der Mathematik,* Vol. 6, pp. 66–76 (1955).

[26] Karzel, H. Verallgemeinerte absolute Geometrien und Lotkerngeometrien. *Archiv der Mathematik* Vol. 6, pp. 284–295 (1955).

[27] Lingenberg, R. Hyperbolisch-metrische Ebenen mit freier Beweglichkeit. *Symposia Mathematika,* Vol. XI, pp. 397–412 (1973), London–New York.

[28] Lingenberg, R. Charakterisierung minkowskischer Bewegungsgruppen durch Beweglichheitsaxiome. *Sitzungsberichte der Bayerischen Akademie der Wissenschaften, mathematisch-naturwissenschaftliche Klasse* pp. 15–34, (1974).

[29] Lingenberg, R. Endliche metrische Ebenen mit Polardreiseit. *Archiv der Mathematik,* Vol. 25, pp. 405–410 (1974).

[30] Lingenberg, R. Endliche minkowskische Ebenen von Charakteristik $\neq 2$. *Journal für die reine und angewandte Mathematik,* Vol. 274/275, pp. 299–309 (1975).

[31] Lingenberg, R. Quadratische Form für S-Gruppenebenen mit vollverbindbaren Punkten. Festband Prof. Dr. H. Lenz, Berlin, 1976.

[32] Lingenberg, R. Parabolische Ebenen. *Beiträge zur geometrischen Algebra,* edited by H. J. Arnold, W. Benz, and H. Wefelscheid, Birkhäuser, Basel, 1977, pp. 217–225.

[33] Lingenberg, R. Vollständige metrische Ebenen. To appear in *Abhandlungen aus dem Mathematischen Seminar der Universität Hamburg.*

[34] Müller, H. Kennzeichnung unitär-minkowskischer Ebenen durch Beweglichkeitsaxiome. Thesis, Karlsruhe, 1974.

[35] Nolte, W. Spiegelungsrelationen in den engeren orthogonalen Gruppen. *Journal für die reine und angewandte Mathematik,* Vol. 273, pp. 150–152 (1975).

[36] Nolte, W. Das Relationenproblem für eine Klasse von Untergruppen orthogonaler Gruppen. Journal für die reine und angewandte Mathematik, Vol. 292, pp. 211–220 (1977).

[37] Nolte, W. Affine Räume als Gruppenräume orthogonaler Gruppen. *Geometriae dedicata,* Vol. 7, pp. 21–35 (1978).

[38] Nolte, W. Projective planes with a collineation group generated by involutions. Preprint 319, Technical University of Darmstadt, 1976.

[39] Nolte, W. Relationen zwischen einfachen Isometrien in orthogonalen Gruppen.

Beiträge zur geometrischen Algebra, edited by H. J. Arnold, W. Benz, and H. Wefelscheid, Birkhäuser, Basel, 1977, pp. 275–278.

[40] Nolte, W. Erweiterte Spiegelungsgruppen metrischer Vektorräume I, II. To appear in *Journal of Geometry.*

[41] Ott, U. Gruppentheoretische Kennzeichnung Pappusscher affiner Ebenen von Charakteristik $\neq 2$. *Journal für die reine und angewandte Mathematik* Vol. 248, pp. 172–185 (1971).

[42] Ott, U. Über eine Klasse endlicher absoluter Geometrie. *Journal of Geometry*, Vol. 1, pp. 41–68 (1971).

[43] Ott, U. Die endlichen Polardreiseitgeometrien. *Journal of Geometry*, Vol. 4, pp. 119–142 (1974).

[44] Ott, U. Über zwei Klassen endlicher S-Gruppen. *Journal für die reine und angewandte Mathematik*, Vol. 297, pp. 1–20 (1978).

[45] Sperner, E. Ein gruppentheoretischer Beweis des Satzes von DESARGUES in der absoluten Axiomatik. *Archiv der Mathematik*, Vol. 5, pp. 458–468 (1954).

[46] Sperner, E. and Ellers, E., see reference 19.

[47] Strubecker, K. Geometrie in einer isotropen Ebene. *Mathematisch-Naturwissenschaftlicher Unterricht* (*MNU*), Vol. 15, pp. 297–306, 343–351, 385–394 (1962).

A more detailed bibliography related to the topic of this book can be found in reference 2 which collects books and papers up to the year 1973. Thus we may list here only publications after 1973. However, some papers mentioned in reference 2 are included, since we referred to some special results contained in these papers without repeating the proof.

INDEX